PRACTICING MATERIALITY

Practicing Materiality

Edited by

RUTH M. VAN DYKE

THE UNIVERSITY OF
ARIZONA PRESS

TUCSON

The University of Arizona Press
www.uapress.arizona.edu

Printed in the United States of America
20   19   18   17   16   15   6   5   4   3   2   1

ISBN-13: 978-0-8165-3127-1 (paper)

Cover designed by Nicole Hayward

Publication of this book is made possible in part by funding from Binghamton University—State University of New York.

Library of Congress Cataloging-in-Publication Data
Practicing materiality / edited by Ruth M. Van Dyke.
       pages cm
    Includes bibliographical references and index.
    ISBN 978-0-8165-3127-1 (pbk. : alk. paper)
    1. Material culture. 2. Anthropology—Methodology. 3. Technology—Anthropological aspects. I. Van Dyke, Ruth M., editor.
    GN406.P73 2015
    306.4'6—dc23
                                    2015005910

♾ This paper meets the requirements of ANSI/NISO Z39.48-1992 (Permanence of Paper).

Contents

PRACTICING MATERIALITY

CHAPTER ONE

Materiality in Practice

An Introduction

Ruth M. Van Dyke

It is little wonder that relationships among things and humans are front-and-center in the contemporary social sciences and liberal arts given the presence of technologies in every conceivable aspect of our (first world) lives. This techno-embeddedness seems to be increasing at an exponential and dizzying pace. Driverless cars ferry Palo Alto software engineers to work. Google Glass connects users to the Internet without the cumbersome interface of a cell phone or tablet screen. Molecular biologists are splicing genomes to "bring back" the extinct passenger pigeon and the woolly mammoth. Hamburgers have been cultured from cow cells, and human organs will soon be grown from stem cells. In less than a decade Twitter and texting have transformed our social landscapes in ways that, I suspect, future scholars will argue rival (or exceed) the changes wrought by the invention of the printing press. Technology has always been ubiquitous in human life, but technological innovations have accelerated to a point where we cannot deny our utter reliance on a vast array of material assemblages that none of us completely understand. I do not refer merely to microchips. Inspired by a character in a Douglas Adams (1992) novel who confessed he had no idea how to build a toaster, design student Thomas Thwaites spent nine months figuring out how to construct a toaster from scratch, from smelting metal and molding plastic through final assembly. Thwaites's (2011) experiment demonstrated just how far removed most of us are from the mundane technologies encountered in daily life, never mind the rocket science.

Clearly this techno-acceleration is having a profound effect on the human experience; in fact, one could argue that technological innovations and the human experience have always been inseparably and mutually co-constructed. To be human without objects is unimaginable. For the past three million years or so human interactions with objects have enabled and constrained our possibilities (Boivin 2008:181–224). For a blind person, the walking cane is an extension of the body; it is integral to the ability to experience the world and to move through it (Merleau-Ponty 1981 [1962]). The laptop and the smart phone are similarly indispensable extensions of (first world) twenty-first-century bodies; without them, society as we know it would not exist. When I unplug from my devices on a week-long rafting trip, I experience a qualitatively different, decelerated daily life, but I am still utterly reliant on a vast assemblage of beings outside of myself—plastic air pumps, butane canisters, rubberized dry bags, carbon alloy paddles, and the helping hands of boat mates when an unexpected snag dumps me into the river.

The materiality of human life has always been inescapable, and perhaps accelerating technologies have always contributed to Western anxieties about their use and control, as depicted by twentieth-century authors and auteurs from Herbert Marcuse (1982 [1941]) to Stanley Kubrick (1968). These anxieties arise within a Western ontology that stretches from the shadowy recesses of Plato's cave through Descartes and Kant to insist that things be kept categorically separate from people. From a Cartesian perspective, things—as categorically different sorts of beings—are the ultimate Other; thus we must find ways to overcome our trepidations and craft uneasy alliances with them. Above all, we must ensure that we keep things under human control.

But an alternate route through this existential distress is being forged by a "posthumanist" or "ontological turn" across the sciences and humanities that is radically reimagining human-object relationships (e.g., Barad 2003, 2007; Bennett 2010; Brown 2004; Bryant et al. 2010; Coole and Frost 2010; DeLanda 2006; Deleuze and Guattari 2007 [1980]; Gell 1992, 1998; Ingold 2000, 2013; Latour 1993, 1999, 2005; Preda 1999; Trentmann 2009). These new perspectives urge us to move beyond Cartesian dualisms and to think of objects and people as ontologically inseparable, entangled within mutually constitutive yet ever-shifting relationships.

Despite a long history of cultural relativism, the embrace of multiple viewpoints, and postcolonial decentering of Western perspectives, anthropologists have been slow to engage with the weightiness and the materiality of objects on their own terms. Miller (2005) suggests that for much of its history, anthropology has tended to trivialize things, treating them as "vulgar material" or relegating them to dependent, epiphenomenal status. Bourdieu's work represents an important shift away from seeing objects as mere resources or symbols. The inseparability of people from their surroundings is integral to Bourdieu's (1977) concept of *habitus*. Building from the phenomenological insights of Merleau-Ponty (1981 [1962]), as well as the political insights of Marx, Bourdieu's theory of practice explores how daily life within constructed material surroundings invents and reinforces worldviews and sociopolitical relations. Foucault's (1977) work on schools and prisons also gives the material a prominent, if not determinate, role in human social lives; but for Foucault, the emphasis remains on the sociopolitical power wielded by those human factions who are behind the materiality of the state.

Over the past decades, many anthropologists working in non-Western contexts have recognized the constructed nature of the Western desire to keep humans and objects categorically separate. Nancy Munn (1977, 1983), for example, describes how objects act as part of distributed persons in Australian *kula* exchange. Marilyn Strathern (1988, 1991) has delineated the concept of the "dividual" in Melanesia where, she argues, there is no ontological or conceptual separation between an individual and "society"; rather, there are ebbs and flows of ongoing exchanges and interactions, and objects are interwoven into these movements and relationships. Working among the Achuar in the Amazon, Eduardo Viveiros de Castro has developed "perspectivism" from the recognition that non-Western ontologies need not categorically separate people from elements of the nonhuman environment, such as animals. Daniel Miller, who uses a dialectical framework to study material consumption within a wide range of contemporary societies, posits that just as we create objects, objects create us through a circular process he calls *objectification* (Miller 1987:19–33, 2005).

Archaeology—the study of the past through the lens of its material residues—would seem particularly well suited to give material objects their due as powerful players within social life. Ironically, however, the marked idealist tendencies of twentieth-century anthropology encouraged many archaeologists to downplay, ignore, and overlook the importance of the material world. Either technologies were considered entirely epiphenomenal to behavior and ideas, or they were seen as the basic determinants of behavior and ideas—in both cases, the material world was a means of access for archaeologists to more interesting dimensions of anthropological inquiry. Hawkes's (1954) ladder of inference expresses these sentiments quite well, with technologies placed firmly on the lowest rung.

An overly simplistic reading of Marx (mode of production determines sociopolitical configuration) led to the technological determinism espoused by White (1949, 1959, 1975) and sometimes unfairly attributed to Steward (1955). For processual archaeologists, objects remained a secondary focus of study, as they sought to understand the "system behind both the Indian and the artifact" (Flannery 1967:120). Experimental archaeologists and ethnoarchaeologists looked into how things were made, but the emphasis remained on the things as a means to the primary goal: reconstructing past human behavior (e.g., Schiffer 1976, 1999; Skibo and Schiffer 2008; Skibo et al. 1999). Postprocessual archaeology actually reified the subject/object and mind/body split still further, as things were considered "symbols in action" or were to be "read" as texts (e.g., Hodder 1982, 1986). The postprocessual focus was less on what people did and more on what they meant, with things pushed still further into the margins.

But with the rise of the posthumanist ontological turn, there has been a general call in archaeology to move beyond Cartesian material/ideal dichotomies and to return materials to a place of primacy in human lives (e.g., Boivin 2008; Henare et al. 2007; Hodder 2012; Knappett and Malafouris 2008; Olsen 2010; Olsen et al. 2012; Shanks 2007; Watts 2013; Webmoor and Witmore 2008). The past five years have seen journals and conference papers burgeoning with references to *networks*, *bundles*, *entanglements*, and *assemblages*, as archaeologists seek new ways of thinking about the relationships among things, places, and people. Some colleagues contend that archaeology may be uniquely positioned to offer insights into object-human relations (Alberti et al. 2011; Olsen 2010).

However, many of these works devote a lot of space to critique and little space to practice. For example, Olsen's (2010) elegant *In Defense of Things* contains compelling arguments, but it is meant as a review of ideas, not as a methodological primer. As Tim Ingold (2007a) has trenchantly observed, there tends to be shockingly little "material" in "materiality" studies (but see Knappett and Malafouris 2008; Watts 2013). All too often, the archaeological reader is left wondering, "so, just how do I operationalize these perspectives in my work?" In this volume nine scholars have taken up the challenge and have developed archaeological case studies that apply perspectives gleaned from the post-Cartesian critique.

In the following section of this chapter, I review the major ideas that have contributed to the materiality literature in anthropological archaeology, and I explore how these works have inspired other archaeologists. But this volume represents more than a series of cases designed to exemplify how theory may be actualized in practice. The authors in this book seek to highlight some of the limitations of each theory and to chart ways that we can move farther. We are particularly concerned that as anthropologists turn toward animals and things, we run the risk of turning away from people, and we abrogate responsibility for intentional action. Thus, by contrast with some of our colleagues, we retain a firmly "anthropocentric" focus, following tracings, linkages, and connections but ultimately contributing to discussions about the negotiation of power within human social relationships.

Entry Points

There are many possible entry points for investigating the potency of material objects and their interactions with humans. Archaeologists frequently have taken inspiration from the work of Bruno Latour and Alfred Gell (although these two scholars do not, in fact, share a synonymous vision of object ontology). Phenomenology—which traces its roots directly to post-Cartesian critique—underlies much of the posthumanist literature and provides a strong set of concepts and methods. Other colleagues are building from indigenous, non-Western perspectives that do not parse the world into living and inert, or material and ideal categories. Daniel Miller and colleagues at University College London (UCL) focus on material culture and consumption from a distinctly Marxist perspective. Peircean semiotics are gaining traction. And, there has been much archaeological discussion about the temporal and spatial connections among objects, places, and people through life histories, enchainments, assemblages, and bundles. In the pages that follow, I review key scholars and concepts that have been influential for the authors in this volume.

Miller and the University College London School

For several decades, archaeologists at UCL have been engaged in studies of "material culture." United by Marxist-inspired perspectives, these colleagues are investigating how aspects of the material world work to enable and constrain relationships among people (e.g., Bender 1998; Buchli 1997, 2002; Miller 1987, 2005; Rowlands 2005). One of the best-known participants in the UCL school is Danny Miller. Originally an archaeologist, Miller has become a leader in studies of contemporary consumer culture. Miller argues for "the humility of things" (Miller 2005:5); objects are important not simply because they constrain and enable, but because their mundane, quotidian status allows them to escape our notice. He argues that studies of mass consumption are necessary not only to appreciate people's relationships to material culture but also to understand social relationships in contemporary industrial societies. A prolific writer and collaborator, his investigations range from automobiles and clothing to media and the Internet (e.g., Horst and Miller 2006, 2012; Küchler and Miller 2005; Miller 1987, 1998, 2001, 2005, 2008, 2010, 2011, 2012; Miller and Woodward 2011).

Miller draws on Hegel, Marx, and Simmel in his concept of objectification—the circular process by which everything we create also creates us (Miller 1987:19–33). It is not just that objects can be agents; it is that practices and their relationships create the appearance of both subjects and objects through the dialectics of objectification, and we need to be able to document how people internalize and then externalize the normative. In short, we need to show how the things that people make, make people (Miller 2005:38). Objectification is a process integral to understanding human social relationships, identities, meanings, and inequalities. Lynn Meskell's (2004, 2005) work on ancient Egypt provides one example of how this might work in archaeological practice. Ancient Egyptians brought the immaterial divine into being through their engagement with monumental or everlasting materials, such as statuary, mummies, and pyramids. Contemporary Westerners in turn consume an idea of the mysterious and vanished Egyptian past through engagement with modern monuments, such as the Luxor hotel and casino in Las Vegas.

Life Histories and Enchainments

The work of Arjun Appadurai on commodities in *The Social Life of Things* (Appadurai 1986) provides another entry point for seeing that the relationships of objects with each other, and with people, are not static. In the same volume, Igor Kopytoff (1986) developed the concept of object life histories to trace the changes that ensue as things move into and out of the commodity state. Walker (1999) and others have picked up on the utility of the life history concept for archaeological artifacts, from their beginnings as raw materials, through production and use, eventual discard, archaeological context, and recovery. Walker and Schiffer (2006) and Schiffer (1999, 2007) expanded upon the life history or object biography approach to develop an explicitly relational schema for people and objects. They see people and objects as equivalent "interactors" within a "cadena," or chain. Sets of people and things intersect along the trajectory of an artifact's life history. Somewhat similarly, Joyce (2003) has employed object life histories to trace interactions between objects and different people across time (see also Mills and Walker's [2008] edited volume, *Memory Work*).

The ideas of *citation* and *enchainment* also describe how objects can reference other objects across time and space. For Derrida (1977), repeated signatures are imperfect citations of one another, each referencing the signatures that have gone before, and presaging those that will follow. Butler (1993) used citation to describe the ways in which genders are performed by imitating recognizable aspects of gender performances that have come before. Archaeologists have employed citation to think about how new objects reference existing materials and relationships, yet transform ideas over time; this is one of the ways in which social memory works (e.g., Jones 2007; Pauketat 2008; Van Dyke 2009). Enchainment expands upon the notion that objects are parts of other entities, so that fragments can reference wholes. For example, Chapman (2004) argues that intentional breakage and deposition of objects linked far-flung aspects of ancient central European societies. Fowler (2004) recognizes dividual, partible, or permeable multi-authored persons.

Latour and Actor-Network Theory

Perhaps the most influential thinker of the posthumanist turn is science and technology studies sociologist Bruno Latour. In *We Have Never Been Modern*, Latour (1993) argues that nature, culture, and discourses about them are ontologically indivisible. By contrast with premodern peoples, modern scholars have attempted to categorically separate nature and culture into discrete subjects studied by specialists. In reality, we cannot and do not treat these phenomena separately; we should admit that "we have never been modern" and think of the world in terms of hybrid interconnections or networks among people, objects, and ideas. Influenced by the work of philosopher Michel Serres, and together with sociologists Michel Callon and John Law, Latour proceeded to develop actor-network theory, or ANT (see especially Callon 1986; Latour 2005; Law and Hassard 1999). Reacting to the social constructivism of the 1980s and 1990s, these scholars seek to completely reshape sociology. In ANT, humans and objects exist on completely equal footing as *actants* in a network. What is important is the connections among the actants, and these connections are ever in flux. Latour charges the sociologist to follow the connections or trace out the billions of ever-changing, myriad linkages among people and things across space and time.

Latour's work has a strong attraction for archaeologists because it fore-grounds not only the material, but also the tracing of connections. Tracing materials across time and space to investigate production, consumption, and social relationships is the very bread and butter of our craft. Follow-ing Latour, Joyce (2008) has argued that a Formative period platform at Los Naranjos in Honduras encouraged the construction of later pyramids by enduring, hosting activities, and reminding people of associated events and meanings. Using a pair of eyeglasses as an example, Webmoor and Witmore (2008) describe a thing as a gathering of technologies, materials, histories, and interactions extending through both space and time. Shanks (2007) and others (e.g., Olsen 2007; Witmore 2007) have followed Latour to advocate for a "symmetrical archaeology" that focuses on relationships between people and things, individuals and structures, past and present.

Peircean Semiotics

Some archaeologists have found inspiration in the semiotics of late nineteenth-century philosopher Charles Sanders Peirce (e.g., Peirce 1992). Although Saussurean semiotics are better known in anthropology, the Sau-ssurean sign/signifier model of meaning reifies a structural understanding of human relationships with objects. Meaning is something that is applied to objects by people; the objects do not exert any reflexive or independent agency in the matter. Peirce, by contrast, developed a more complicated and dialogic model in which things contribute to subjects' understand-ings of them. Peirce's tripartite, relational semiotics include the sign itself (which is often but not necessarily material), the object or idea to which the sign refers, and the sense made of the sign (interpretation). Signs can refer to a corresponding referent through iconicity (resemblance, as in the picture of a small house on the computer screen means home page), in-dexicality (a direct relationship, as in smoke means fire), and convention (accepted symbolism, as in the Union Jack means the British Empire). The material world, and the changing senses people make of it, are bound up in interesting ways in this perspective, allowing us a way to transcend discursive/nondiscursive and subject/object separations. Some archaeolo-gists thus see great potential in adopting a Peircean approach (Crossland 2009, 2013; Jones 2007; Joyce 2008; Liebmann 2008; Preucel 2010).

Robert Preucel (2010) offers a comprehensive exegesis of Peirce's ideas and their potential utility in archaeology. Preucel examines the history of twentieth-century archaeological theory and explains how processual, postprocessual, and cognitive archaeologies all allow us to capture only part of what is relevant about past relationships between thought and material. He contends that Peircean semiotics can provide a more interesting way to think about how people, materials, meanings, and experiences intersect. Preucel provides examples drawn from the seventeenth-century Pueblo Revolt in the American Southwest and from his work at Brook Farm, a nineteenth-century Fourier settlement in Massachusetts. Zoë Crossland similarly advocates for a Peircean framework in archaeology. She has employed a Peircean semiotic approach to think about the afterlives of dead bodies in forensic anthropology (Crossland 2009) and to analyze a nineteenth-century missionary movement in southern Africa (Crossland 2013).

Bundles, Entanglements, Assemblages, and Meshworks

Because relationships necessarily imply the presence of multiple elements, archaeologists are creatively examining sets of disparate things, people, feelings, and ideas that are linked by some kind of association. Some scholars have advanced the term "assemblages" for these groupings (following Deleuze and Guattari 2007; see also DeLanda 2006), while others prefer "meshworks" (Ingold 2011), bundles (Keane 2003, 2005), or entanglements (Hodder 2011, 2012).

The concept of the bundle—derived by Webb Keane (2003, 2005) within a Peircean semiotic framework—has been particularly influential for the authors in this volume. Keane (2005:187) illustrates how the material is inextricably bound up with multiple qualities and meanings with an example from a children's book: a little girl "likes red," but an adult points out that a gift to the child cannot be a color; it must be attached to "something red." In this example, presumably a gift of "red" could be an apple, a pair of ruby slippers, or a toy fire engine; these very different objects would carry the same import to the little girl. But in any of these examples, redness would be accompanied by other sets of tactile attributes, such as crunchiness, smell, shininess, flexibility, or potential to roll (or not). Emotions, memories, and other associations would likely also be present. Keane refers to this copresence of multiple attributes as bundling.

Bundles incorporate not only objects and people, but assumptions, knowledge, meanings, and ideas. Zedeño's (2008, 2013) discussion of Blackfoot medicine bundles is a particularly concrete example, but bundles need not necessarily take discrete physical form. Pauketat (2013) employed the concept of bundling to think about relationships among objects, people, and places. The flexible metaphor of the bundle describes how elements can be grouped together, loosely or tightly, in ever-shifting relationships, and across both space and time. Bundling integrates emotional or sensuous qualities, such as color, texture, nostalgia, or repulsion.

In *Entangled*, Ian Hodder (2011, 2012) adopts the trope of entanglement, rather than the bundle, to somewhat similar effect. He describes the complicated relationships between humans and things, and he argues that we are entrapped in the world of things because we develop obligations to them, and they to us. He recognizes that the Neolithic occupants of Çatalhöyük were able to construct certain kinds of modular houses because suitable mudbrick clays were available . . . but the houses would not stay up without constant maintenance by the humans, drawing the two together in an interdependent relationship. Replastering, clay pots, earthen figurines, and fertile soil were all part of the assemblage or bundle of humans/farming/sedentism/earth at Çatalhöyük.

Deleuze and Guattari (2007 [1980]) develop the trope of the rhizome (as opposed to the rooted tree or fragmented radicle) to illustrate the world as interconnected, irreducible, ebbing, and flowing tracings of interconnections. Grass provides one evocative rhizomatic metaphor; plateaus (which are always between other places) provide another. Deleuze and Guattari urge us to think in terms of assemblages—entities made up of interconnections and fragments of the material, the social, and the semiotic. In *Lines*, Ingold (2007b) takes inspiration from Deleuze and Guattari's idea of assemblages, but he prefers the concept of a meshwork to describe the rhizomatic, moving, living, co-creative entanglements of humans and other beings.

Building from Deleuze and Guattari's ideas, DeLanda (2006) explores how assemblages might help us understand multiscalar, flexible relationships among individuals, communities, and nations. Assemblages are held together by territorializing processes and pulled apart by deterritorializing processes. Assemblages are wholes that cannot be reduced to parts because the whole is not an aggregation of parts; it consists of the contingent relationships among them. Harris (2012) uses DeLanda's assemblage concept to discuss multiscalar relationships in Neolithic and Bronze Age British communities.

Phenomenology

The tactile, experiential qualities of materials are clearly part of the conversation, and this leads us to consider phenomenology. Phenomenology in archaeology derives from existential philosophy (see Casey [1996], Gosden [1994], and Thomas [1996] for good overviews). Existentialism posits that we bridge the distance between objective reality and our subjective consciousness through existence or daily experiences and participation in the world. Phenomenology emerged in the twentieth century through the works of Edmund Husserl (1960 [1931]), Martin Heidegger (1962 [1927]), and Maurice Merleau-Ponty (1981 [1962]). Husserl argued that we know the world not through pondering it (à la Descartes) but rather through daily life, experience, and perception. Conscious, intentional, bodily engagement with the physical world is the starting point for all knowledge. For Husserl, an idealist, understanding the world hinges around conscious perceptions of experience. Heidegger, a student of Husserl, moved the conversation away from epistemology—the study of how we know what we know—toward ontology—the study of being, or existence. Like Husserl, Heidegger saw the human body as the point of dialectical mediation between consciousness and the physical world, subject, and object. However, for Heidegger, bodily experience has precedence over ideals and intentions, because experience creates the context in which all else takes place. Existence involves a mesh of contingent relationships with other beings and objects in both spatial and temporal dimensions. This is explored in Heidegger's concept of *Dasein*, translated as being-in-the-world, or dwelling. Maurice Merleau-Ponty (1981 [1962]) further elaborated upon these ideas and focused on the intersections between the body, perception, consciousness, and place.

Heideggerian phenomenology thus offers one route for transcending the artificial and categorical divide. All knowledge stems from our bodies—which are themselves material. Objects are not categorically distinct but rather are part of this world, extensions of ourselves and our experiences, coming into or out of focus in the course of our bodily interactions. Heidegger (1962 [1927]), for example, famously discusses the hammer—a common household tool that usually is not part of daily experience. However, if a person needs to drive a nail and does not have a hammer "ready to hand," then the object makes itself felt—it comes into focus—by its absence.

Although phenomenology in archaeology is best known through the landscape work of Tilley and his followers (Cummings and Whittle 2004; Tilley 1994, 2004; Van Dyke 2007), these philosophical ideas have much wider epistemological relevance in archaeology. Thomas (1996, 2004), for example, employs Heidegger in a sophisticated critique of archaeological practices that are, he contends, mired in modernist, Descartian dichotomies. Thomas advocates Heidegger's concept of Dasein, or being-in-the-world, to break down the false separation between human existence and the material. A phenomenological perspective encourages us to think of objects, meanings, places, and people as continually, mutually constitutive, flowing into one another across time. For example, rather than a transcendental expression of structural meaning, a Neolithic chambered tomb can be seen as one in a chain of performative events linked with recent and older ideas and practices.

Gell: Enchantment, Secondary Agency, and Extended Minds

Phenomenology figures largely in the work of British anthropologist Alfred Gell. Gell sought to develop a theory of art in which objects are active participants, not merely passive signifiers. He was interested in the ways in which complicated visual creations, such as kula canoe prows or Asian mandalas, can draw in the eye and "enchant" the viewer (Gell 1992). And, to those without specialized knowledge, the technological creation of complex objects appears as mysteriously enchanting. In his posthumous work *Art and Agency*, Gell (1998) created an elegant, complicated schema to describe different kinds of relationships among artists, objects, and meanings. The schema rests upon the ideas of abduction (an inferred relationship) of the index (from Peircean semiotics). We commonly think of artists as active, or *agents*, and the objects they create as passive, or *patients*. Gell pointed out that these relationships are much more complex, involving references to ideal prototypes; objects can also act as agents on human patients. Gell developed a *nexus*, which illustrates all the various possible agent and patient relationships among prototypes, artists, indexes, and recipients. Gell provided an innovative perspective on the ways objects can actively construct relationships and meanings, but (by contrast with Latour) he was careful to distinguish between the *secondary agency* of objects and the *primary agency* of humans with consciousness and intentionality.

Gell also was interested in how artistic and technological styles extend across time and space. Following Husserl, as well as sociocultural anthropologists like Strathern and Munn, Gell argued that persons are not biologically bounded entities but are distributed by means of materials across space and time. Gell's concept of *extended mind* describes how the external movements of objects (e.g., kula shells) are also part of the internal workings of humans (e.g., the kula operator's knowledge of the game). Gell's depiction of the slight changes in Maori meetinghouses across time is essentially a discussion of the motivations behind stylistic seriation. Each subsequent house resembles one that came before and thus is recognizable, but each house is also subtly different, and eventually this results in change. These aspects of his work have obvious utility for archaeologists. Gosden (2005), for example, draws upon Gell's notion of secondary agency to describe how repetitive human actions create assemblages of objects—such as bronze fibulae—in Roman Gaul. Once these assemblages were brought into existence, they in turn affected people's actions (perceived as signifying "Roman-ness").

Ingold: Taskscapes and Dwelling

Phenomenology has contributed to the thought and work of sociocultural anthropologist Tim Ingold over the past two decades (Ingold 1993, 2000, 2007b, 2011, 2013). Beginning with the Heideggerian notion of *dwelling-in-the-world*, Ingold (1993) points out that people are never disembodied but move seamlessly through the world and through time. Meaning is created and sustained through crosscutting fields of relationships among people, objects, and places that gradually unfold. *Tasks*—the practical activities carried out by people on a daily basis—are the constitutive acts of dwelling. The *taskscape*, then, is an inherently social array of spatially and temporally related activities. Ingold's concept of the taskscape adds temporality and practice to the spatial study of landscape.

Ingold's (2000) collection of essays entitled *The Perception of the Environment* is a groundbreaking work of transdisciplinary exploration that draws from phenomenology, anthropology, biology, and ecological psychology. The essays, organized around livelihood, dwelling, and skill explore how humans, animals, and things coconstruct one another through their mutual perceptions and experiences. Ingold continues to carry phenomenological ideas forward in innovative directions. In *Lines*, Ingold (2007b) engages in a widely ranging discussion organized around the contrast between the modernist vision of the world as disconnected points or separate categories, and an antimodern or non-Western vision of the world as continuous, curved, lived, and open ended. Modernity separates journeys (lines) into series of discrete places (dots) (Ingold 2007b:75). By contrast, Ingold takes inspiration from phenomenologists Merleau-Ponty (1981 [1962]) and de Certeau (1984) to advocate wayfaring, moving across the earth, with the journey itself as the focus, paying attention to what you are seeing and doing. Ingold further develops his anti-essentialist, transdisciplinary perspectives in *Being Alive* (2011), and a second collection of essays entitled, *Making* (2013).

Animals and Animisms

Science and technology scholar Donna Haraway has long advocated for eliminating boundaries between humans and nonhumans. In her "Cyborg Manifestos" (1985, 1991), both hailed and reviled by third-wave feminists, she critiqued the essentialized Judeo-Christian view of "woman" and advocated for moving beyond gender altogether into an emancipatory realm of human-machine hybrids. Haraway's more recent (2003, 2008) work deals with the interrelationships among humans and animals, as she challenges a modernist vision of humans as discrete enemies and animals as Other, portraying the "face-to-face" relationships of dogs and people as knots of connections that bring into focus both entities' complex histories encompassing ecological, evolutionary, historical, and political dimensions.

In *Vibrant Matter*, political scientist Jane Bennett (2010) similarly advocates for transcending any separation between human, animal, and material worlds. She sees *vital materiality* as a quality present across all bodies. Looking toward ecology, she argues that we need to recognize the agency of nonhuman forces in complex contemporary situations, such as electricity blackouts, chemically leaching landfills, and stem cell research. We should "give up the futile attempt to disentangle the human from the nonhuman," Bennett (2010:116) argues, and instead try to build a better and more "civil" world that includes assemblages of humans and nonhumans.

Indigenous ontologies need not and generally do not involve a categorical division between nature and culture. Clearly, non-Western and past peoples are unlikely to have viewed the world using Western, Cartesian dualisms. Here the work of Amazonian anthropologists Philippe Descola (1994) and Eduardo Viveiros de Castro (1998) has been particularly influential. Descola sets up a four-part schema to describe possible views regarding humans and animals. *Analogism* sees humans and animals as completely different in form and being. *Totemism* considers groups of animals and humans as sharing features that set them apart from other groups. *Naturalism* describes the Western, ontological nature/culture divide in which humans and animals share biology, but human consciousness sets us apart. *Animism*, by contrast, involves a continuity between humans and animals; humans have outwardly different forms but share characteristics attributed to animal species.

More than a half century ago, Jacob von Uexküll (2010 [1934, 1940]) explored the idea that humans and animals perceive the world differently by virtue of our different bodies. In a famous example he pointed out that light, color, sounds, and other familiar aspects of the human sensorium are not part of the experience of the tick, which is concerned only with body heat and blood. Von Uexküll's tick, not surprisingly, is cited by phenomenologists to illustrate the importance of bodily experience in constructing reality. Viveiros de Castro took this still farther with his concept of Amerindian *perspectivism*. Viveiros de Castro worked in the Amazon with the Achuar, who believe that animals and plants possess human souls inside bodies that merely look nonhuman on the outside; thus, humans, plants, and animals are ontologically indistinguishable and must be treated similarly. Like von Uexküll, Viveiros de Castro contends that the body one occupies creates the subject's perspective; animals, therefore, see the world differently. But, animals and humans are all beings with spiritual, as well as corporeal, aspects, so there is no need to artificially separate them.

If we put aside the separation of humans and animals, or living and inert beings, as particularly Western, modern constructions, we can recognize other perspectives in which objects and animals are active and animate participants in the human world (e.g., Bird-Davis 1999; Brown and Walker 2008; Watts 2013). Mills and Ferguson (2008), for example, argue that shell trumpets were considered as animate objects in the ancient American Southwest. Zedeño (2008, 2013) explores the agency and animacy of Blackfoot medicine bundles as a key part of the Blackfoot worldview. In Blackfoot ontologies, the bundles are living object-persons that transfer power from objects to people and vice versa. Melanesian *malanggan* (wooden mortuary carvings) are another well-studied group of objects that are animated, and de-animated, as part of the process of commemoration and forgetting (Küchler 2002). Şule Can (this volume) points out, however, that it is just as dangerous to generalize in non-Western as in Western contexts. For example, for Alawite followers of Islam, objects may have resonant, affective relationships with humans, but it would be considered blasphemous to project such ideas onto animals.

Summary

This cursory overview does not by any means exhaust the breadth and depth of posthumanist, post-Cartesian, or materialist thought in the contemporary humanities and social sciences. Rather, I have attempted here to provide thumbnail sketches of ideas that have been particularly influential for archaeologists. The scholars whose work I have reviewed are building from diverse literatures and working toward different goals, but they share the position that nonhumans (objects, materials, and animals) should be elevated from the epiphenomenal, secondary, or invisible states in which they have languished in modern Western scholarship. And there are other strong common threads that run through these authors' ideas. One is a vehement anti-essentialism, or a desire to transcend Cartesian dualisms, such as subject/object, human/nonhuman, and culture/nature. Another is an emphasis on connections rather than static entities, lines rather than points, and networks rather than nodes. The connections or relationships form the primary focus of study—not the entities or end-points. These connections or relationships are transitory, momentary, and flexible across time and space. And, they are not necessarily hierarchical but are tangled and complex; they resemble rhizomes as opposed to roots. And, finally, whether or not each of these scholars would label him/herself a phenomenologist, there is a strong current of phenomenological episte-mology in this literature. To be alive is to interact with the tactile, sensual, meaningful, material world around us. We know the world through our embodied experiences of it.

Posthumanist vantage points can be empowering for anthropologists as they challenge us to think outside the Western box. They can encourage us to find ways to more faithfully represent and engage with non-Western perspectives, past and present. If we dissolve the ontological divide be-tween people and things, there will be less to fear from accelerating tech-nology. Some scholars (e.g., Bennett 2010; Haraway 2003) look forward to a new age of heightened equality for all peoples, animals, and things. But in many cases, the elevation of things seems to come with what I would argue is a steep price tag—a loss of focus on people.

In Defense of Anthropocentrism

As suggested by the label *posthumanism*, it is becoming commonplace to see calls in archaeology for the discipline to become less "anthropocen-tric." In ANT, as in symmetrical archaeology, people, objects, and linkages in a network must be seen as flat and neutral so as to avoid reifying any preconceived ideas of hierarchy or structure. Authors taking inspiration from Latour (e.g., Harris 2012; Knappett and Malafouris 2008; Overton and Hamilakis 2013) insist that objects and animals cannot be relegated to secondary or reflective status but must be considered as equal players alongside humans in decisions, activities, abilities, and beliefs. "Things are us!" as Webmoor and Witmore (2008:61) proclaim. This is meant to be emancipatory, as we are freed from the flawed idealism of modern-ist thinking and move into a nonmodern, less hypocritical age (Latour

1993). "All bodies become more than mere objects, as the thing-powers of resistance and protean agency are brought into sharper relief" (Bennett 2010:13). Bennett holds that the changes brought by a brave, new, "vibrant materialism" will, in hindsight, resemble the passage of civil rights legislation or women's suffrage (Bennett 2010:109).

The authors in this volume have at least two serious objections to this extreme position. First, we disagree that it is productive or necessary to shift our attention away from people; in fact, we see it as quite dangerous. We are, after all, anthropologists, not geologists or thingologists—shouldn't the social remain at the center of what we do (Fowles 2010; Starzmann 2013)? As anthropologists, we are not merely engaged in a philosophical exercise. We would like to see our work advance the causes of social justice in our world. If "things are us," then humans must also be considered things . . . a stance that would allow and even encourage us to think of humans as commodities (see Hauser, this volume).

This plaint carries no weight with the nonanthropocentrists, apparently. Bennett (2010) sees posthumanism as offering marginalized peoples a "safety net"; if all humans and objects are equivalent, no humans or objects are elevated above others. But it is unclear how this philosophical position translates into the real world where people (and objects?) continue to be privileged and suppressed, exalted and excluded. For Latour (2005:63–64, 85), a move away from social justice is a necessary step; only after we have finished tracing our network can we consider power (Latour 2005:248–253). He urges us to overcome our "infatuation with emancipation politics" (Latour 2005:59) and clearly stakes out his anti (post?) -Marxist agenda when he quips, "social scientists have *transformed* the world in various ways; the point, however, is to *interpret* it" (Latour 2005:42). But Winner contends that posthumanist "interpretive flexibility soon becomes moral and political indifference" (Winner 1993:445). The tracing of networks is a painstaking, laborious, and potentially infinite task . . . and while social scientists are tracing and describing, people are suffering. Philosophical debates aside, a critical, biological difference between objects and humans is that objects lack a central nervous system (Ingold 2008). Objects cannot suffer, whereas humans can and do.

Our second objection involves the posthumanist discard of the thorny issue of intentionality. Posthumanist perspectives appear attractive in part because they provide us with a way out of the intentionality conundrum that bedevils all social scientists: to wit, how do intentions, individuals, collectivities, the exercise of will, resistance, and unintentional consequences interplay to shape social change? Clearly, actions are not intentions (Giddens 1984:9–12), just as agency is not consciousness (Ingold 2008:213–214). No actor is all-powerful or omniscient. Chaos theory and the "butterfly effect," not to mention observations of daily life, tell us that the relationships between actions and intentions are far more complex than *actor + intention = result*. Thus, clearly not all change is caused by the strategic actions of omniscient agents; some (Pauketat 2000) might even argue that no change is caused this way. So how do we disentangle or follow threads of intentions through the complex meshwork of the past? With a flat ontology, if we bring materials into the picture as "agents" (like Formative period monuments [Joyce 2008], or Salado polychrome bowls [Walker and Burt 2009]), we can give everything (or nothing) the same level of intentionality. We can relinquish as irrelevant the problems of who caused what. Everything causes everything in an endless meshwork of reactions and actions. As Webmoor and Witmore point out, there is relief in being able to "shed the belief that the initiative always comes from the thinking, free-standing human being" (2008:59).

However, if we set aside intentionality, we open the door for a serious ethical problem. If intentionality no longer matters, then people with evil intentions cannot be held accountable for their actions (Arendt 2011 [1953]). Latour (1994) famously argues that the gun does not commit murder, nor does the man, but the gun and man together as a hybrid entity, thus shifting responsibility for the murder away from the man and to the gunman. But does it matter to the murder victim whether the gun and the man acted as an ontological assemblage? In too many recent, tragic events it is clear that easy access to guns can facilitate violence. But guns, as objects, cannot be imprisoned or punished for these crimes. The philosophical discussion about the messiness of intentionality once again distracts us from working to prevent real human suffering.

Therefore, the authors in this volume remain firmly and unapologetically "anthropocentric," in the sense that we are anthropologists ultimately interested in the relevance of our work for human beings, not for objects.

Practicing Materiality

The most compelling aspect of the "materiality turn" in archaeology is the exhortation for us to put the material back into our social analyses—to give materials the front-and-center roles that they actually play in human lives. The authors in this volume have begun to think of objects as active players in human life, rather than simply as passive symbols or means to other ends (e.g., subsistence or social power). Rather, we see objects as part of a lived aesthetic continuum that affects, enables, and constrains meaning and behavior. We have focused our attention on mutually constitutive relationships rather than static categories. And, where relevant, we have attempted to think through non-Western perspectives.

All of the authors in this volume adopt an explicitly phenomenological stance. These anthropologists are striving to understand objects' powers to interact with us, not as sentient beings, but as the foci of meanings and notions that we sense and project, and that materials reflect, distort, and transform. We seek a tactile, sensuous engagement with the world of the material—one that recognizes the liminal, the emotional, and the extraordinary enchantment of technology. Our relationships with materials are neither strictly deductive nor inductive, but, as Gell suggests, abductive. Places and things from the past carry a sort of "intraworldliness"; they once were part of a different social life but are now part of ours (Thomas 2006).

Gell's notions of secondary agency, the abduction of the index, and the extended mind also have been particularly influential for authors in this volume, particularly in the work of Şule Can, Rui Gomes Coelho, and Brittany Fullen. Şule Can, a sociocultural anthropologist, writes about the Nusayri, an Arab Alawite Islamic sect in Turkey, whose rituals are entangled with *zyaras* (sacred places) and objects. Zyaras are localities that have become sacred by association with saints or revered religious leaders. They exist as physical buildings tended by communities of believers who leave, use, and take away objects from the zyara as part of religious practice. Can describes the constructive role zyaras and associated collective memory, ritual practices, and beliefs have for Arab Alawite identity. Gell's concept of the art nexus helps Can untangle the distributed agency and mutual coconstruction of saints, zyaras, objects such as healing stones, and communities of believers. Ultimately, however, Gell's nexus falls short of illustrating what is for Can one of the most significant dimensions of zyaras—their role as focal points for Arab Alawite political and religious struggle. To be Nusayri is to practice one's religion, and to practice one's religion is to have access to zyara; but zyara cannot be moved and thus are unavailable for Arab Alawites in diaspora.

Rui Gomes Coelho focuses on ceramics dating to the rise of the Portuguese Empire between the sixteenth and eighteenth centuries. Coelho combines phenomenological insights with Gell's notion of extended mind to describe changes in ceramic assemblages across time and space during a period when concepts about the body and the state were undergoing dramatic transformation. At a site in Setúbal on the Portuguese coast, late sixteenth- and early seventeenth-century archaeological contexts were dominated by earthenwares, but late seventeenth- and early eighteenth-century contexts contained increasing frequencies of faience and porcelain. These changes are not simple reflections of shifting fashions or expanding economic access to goods. Rather, Coelho describes how before the rise of the empire, within a world dominated by Catholic morality, earthenwares were bound up with ideas about humility, virtue, the fragility of human life, and sensory engagements with clay. With the rise of the empire came different ideas about the relationship of the body to the state, and a new set of material associations. Blue and white faience and porcelain connected Portugal with the idea of empire as the Portuguese encountered it in China. More sterile and less chemically volatile faience and porcelain vessels disciplined the senses just as bodies were transformed into imperial subjects. Extended mind helps Coelho to trace sensory and symbolic relationships among people and ceramics over time at Setúbal, and over space as Portugal became a global power. Ultimately, however, Coelho's analysis moves beyond these Gellian insights to address how ceramic objects helped constitute and create an early modern state characterized by bodily and territorial control.

Brittany Fullen engages with a somewhat similar set of issues, as she uses Gell's idea of secondary agency to assess the distribution of Huamanga ceramics dating to the Andean Middle Horizon. This polychrome style, found throughout the expanding Wari Empire of highland Peru, is closely associated with Huari (the Wari capital). Fullen begins her analysis from the perspective of the Huamanga pots—quotidian bowls and cups commonly found in feasting contexts at the secondary site of Conchopata. She argues that Huamanga design motifs were flexible prototypes that exerted influence on the people who made and used the pots. Some Huamanga ceramics were meant to carry Wari identities outward, but Huamanga vessels also were replicated, reinterpreted, and repurposed by local communities within processes of assimilation, as well as, possibly, resistance.

Bundling is another powerful trope employed by many of the volume's authors, particularly Erina Gruner, Tanya Chiykowski, Halona Young-Wolfe, and Jessica Santos. Gruner looks at bundles of ritual objects, and Santos develops the trope of a *protest bundle*. Chiykowski combines the idea of a pot as a bundle with an object biography or life history approach derived from Kopytoff (1986). Young-Wolfe uses insights from phenomenology and ecological psychology in her study of another kind of physical bundle—monuments built of net bags filled with stones.

Erina Gruner uses the concept of the bundle to look at curated assemblages of ritual paraphernalia in ancient and historic Ancient Pueblo societies in the American Southwest. She argues that bundled ritual assemblages found in Chaco Canyon and the Flagstaff area enchained objects, people, and practices across time and space and transferred practical knowledge about the bundles' construction, use, and care. Because they are part of exclusive sacred knowledge, ritual bundles contributed to the rise of elite power. When bundles were cached or obliterated, religious knowledge and identity was similarly dissolved and dispersed.

Tanya Chiykowski combines bundling and object biography to trace the materials, qualities, and social relationships found in plainware ceramics from the Trincheras tradition in northern Mexico. Building from a deep and broad ethnographic tradition in the southwestern United States, Chiykowski describes the life history of a quotidian pot, from the gathering of clay, to the shaping and decoration of the vessel, through firing, use, breakage, and discard, to collection and curation as an archaeological artifact. As Boivin (2008), Hodder (2012), and others have observed, clay is a tactile, fictile material with interesting transformative properties that were likely significant for ancient potters. Chiykowski argues that Trincheras potters saw plainware vessels as animate beings. The malleable clay, like all life, comes from the earth. Pots become more fragile and more durable through the firing process, and they "cry" and "die" when they shatter and break, only to be resurrected in afterlives as artifacts in collections and subjects of controversy over cultural affiliation.

Like Tim Ingold (2000), Halona Young-Wolfe sees resonance between Heideggerian phenomenology and ecological psychology. Heidegger's being-in-the-world draws our attention to quotidian practices of perception—seeing, touching, looking, and doing. Ecological psychologists (e.g., Gibson 1979) have demonstrated that perception does not happen internally, "in the mind" but rather takes place through the process of interacting with the material environment. When cognition is conceived as an interactive, ongoing process, there can be no subject/object divide; instead, perception, living, dwelling, and being are all temporally and material situated (following Ingold's [1993] concept of taskscape). Young-Wolfe employs these insights to examine issues surrounding *shicra*, a material employed in the emergence of monumental architecture in Late Archaic coastal Peru. Shicra—net bags made of local sedges and filled with earth or stone—are used as internal fill in the monuments. Other Andeanists have focused on the possible functions of shicra within the larger context of the monuments (earthquake protection, labor measurements, and ritual demarcation). By contrast, Young-Wolfe thinks about the tactile experiences of making and using shicra. The strong and flexible shicra resonate with a host of other material dimensions of daily life—burial mats, carrying baskets, and fishing nets. Shicra catch and contain the earth just as Archaic Andeans had long caught and contained plants, fish, and animals and just as the new monuments now catch and contain the shicra. Young-Wolfe's case illustrates how monuments and larger-scale sociopolitical organization can emerge out of continuities involving daily, familiar taskscapes.

Contra Latour (2005:42), the authors in this volume do not think we should abandon our attempts to transform (rather than merely interpret) the world. Accordingly, Jessica Santos's case study is an overtly political and potentially transformative sociocultural analysis of the involvement of undocumented immigrants in contemporary public protests. Using media images, as well as blog excerpts, articles, and personal interviews, Santos investigates the intersections among bodies, objects, and public spaces that characterized two major public protests—the immigration reform demonstrations in 2006 and the Occupy Wall Street movement in 2011. Santos develops the trope of the protest bundle as a way to trace the reflexive, tangled interactions of people, places, practices, and things that characterized these movements. When undocumented immigrants, who in daily life hide in the shadows, begin to move collectively in public spaces, they transform themselves from "bare life" (people without rights or citizenship, sensu Agamben [1995]) into political subjects with voices. Protest bundles help describe how those with voices and those without can work to change society through collective action.

The volume concludes with a review and discussion by Mark Hauser, whose own archaeological work on slavery in the Caribbean provides a poignant reminder of the potential risks in conflating people and objects. Hauser frames the discussion in terms of materiality as "problem space." He reviews and reiterates the shared ways in which each author—through case studies focused around some quite active objects—returns to primary concerns of power relations among people.

Materiality in Practice

The authors in this volume make contributions to both method and theory and illustrate how posthumanist frameworks can be employed in analyses while simultaneously highlighting these frameworks' boundaries and limitations. Many anthropologists are creating new *ideal* models or structures for *thinking about* materials, in which the materials under discussion seem to be primarily placeholders for ideas. But this merely re-reifies the material/ideal split. Admittedly, it is difficult (perhaps impossible) to discuss anything without creating labels. Latour (2005:141–156) tells us that the way around this problem is to simply describe the linkages among people and objects, but he offers no guidance as to how the scholar knows when to move from description to explanation. We offer the case studies in this volume to try to move the conversation forward, grappling with the very difficult methodological challenges of the post-Cartesian critique. As Miller (2005:43–44) advises, "to conduct anthropology we need to hitch the (philosophical) wheel back to a vehicle that returns us to the muddy paths of diverse humanity."

The strengths of the case studies contained in this volume lie in their power to get us to think in new ways about the relationships among the social, bodily, and material dimensions of our world—in short, to practice materiality. It is in the process of operationalizing these theories that we can more clearly see their limitations and attempt to transcend them. In our view, practice must also involve praxis—engagement with the politically and social relevant inequalities that are enabled and constrained within networks, meshworks, and bundles. Thus, all of the cases in this volume highlight the ways in which human-thing relationships are part of historically grounded negotiations of social and political power. Our engagements in these chapters are with plainwares and faience, healing stones and net bags, and swallowing sticks and microphones, but our concerns are ultimately with the relevance of these bundled assemblages for human experiences.

Acknowledgments

This volume would not exist without the hard work and long-term engagement of an extraordinary group of young scholars—Şule Can, Tanya Chiykowski, Rui Gomes Coelho, Andrea Fink, Brittany Fullen, Erina Gruner, Torin Rozell, Jessica Santos, and Halona Young-Wolfe. I thank all of them for the discussions, debates, and dialogues that led to this volume. Along the way, I have learned from conversations with Reinhard Bernbeck, Severin Fowles, Deniz Kahraman, Randy McGuire, Matthew Palus, Tim Pauketat, Michael Shanks, Maresi Starzmann, Angela Vandenbroek, Mary Weismantel, and Martin Wobst. I appreciate constructive feedback on this manuscript offered by Lars Fogelin, Mark Hauser, Josh Reno, and two anonymous reviewers. My thoughts on materiality in archaeology continue to evolve, but I take responsibility for any omissions, as well as errors, in representation or interpretation to be found in this work.

References Cited

Adams, Douglas. 1992. *Mostly Harmless: The Fifth Book in the Increasingly Inaccurately Named Hitchhiker's Trilogy.* Harmony Books, New York.

Agamben, Giorgio. 1995. *Homo Sacer: Sovereign Power and Bare Life.* Stanford University Press, Palo Alto, California.

Alberti, Benjamin, Severin Fowles, Martin Holbraad, Yvonne Marshall, and Christopher Witmore. 2011. Worlds Otherwise: Archaeology, Anthropology, and Ontological Difference. *Current Anthropology* 52(6):896–912.

Appadurai, Arjun. 1986. Introduction: Commodities and the Politics of Value. In *The Social Life of Things*, edited by Arjun Appadurai, pp. 1–63. Cambridge University Press, Cambridge.

Arendt, Hannah. 2011 [1953]. On the Nature of Totalitarianism: An Essay in Understanding. In *Essays in Understanding, 1930–1954: Formation, Exile, and Totalitarianism*, edited by Hannah Arendt, pp. 328–360. Random House, New York.

Barad, Karen. 2003. Posthumanist Performativity: Towards an Understanding of How Matter Comes to Matter. *Signs* 28(3):801–831.

———. 2007. Meeting the Universe Halfway: Quantum Physics and the Entanglement of Matter and Meaning. Duke University Press, Durham, North Carolina.

Bender, Barbara. 1998. *Stonehenge: Making Space.* Berg, Oxford and New York.

Bennett, Jane. 2010. *Vibrant Matter: A Political Ecology of Things.* Duke University Press, Durham, North Carolina.

Bird-Davis, Nurit. 1999. "Animism" Revisited: Personhood, Environment, and Relational Epistemology. *Current Anthropology* 40(S1, Special Issue: Culture—A Second Chance?):S67–S91.

Boivin, Nicole. 2008. *Material Cultures, Material Minds: The Impact of Things on Human Thoughts, Society, and Evolution.* Cambridge University Press, Cambridge.

Bourdieu, Pierre. 1977. *Outline of a Theory of Practice.* Cambridge University Press, Cambridge.

Brown, Bill. 2004. Thing Theory. In *Things*, edited by Bill Brown, pp. 1–16. University of Chicago Press, Chicago.

Brown, Linda A., and William H. Walker. 2008. Prologue: Archaeology, Animism and Non-Human Agents. *Journal of Archaeological Method and Theory* 15(4):297–299.

Bryant, Levi, Nick Smicek, and Graham Harman, (editors). 2010. *The Speculative Turn: Continental Materialism and Realism.* RE Press, Melbourne.

Buchli, Victor. 1997. Kruschchev, Modernism, and the Fight Against *Petit-Bourgeouis* Consciousness in the Soviet Home. *Journal of Design History* 10(2):187–202.

Buchli, Victor, (editor). 2002. *The Material Culture Reader.* Berg, Oxford and New York.

Butler, Judith. 1993. *Bodies That Matter.* Routledge, New York.

Callon, Michel. 1986. Some Elements of a Sociology of Translation: Domestication of the Scallops and the Fishermen of St Brieuc Bay. In *Power, Action and Belief: A New Sociology of Knowledge*, edited by John Law, pp. 196–223. Routledge and Kegan Paul, London.

Casey, Edward S. 1996. How to Get from Space to Place in a Fairly Short Stretch of Time: Phenomenological Prolegomena. In *Senses of Place*, edited by Steven Feld and Keith H. Basso, pp. 13–52. School of American Research Press, Santa Fe.

Chapman, John. 2004. *Fragmentation in Archaeology: People, Places, and Broken Objects in the Prehistory of South-eastern Europe*. Routledge, London.

Coole, Diana, and Samantha Frost, (editors). 2010. *New Materialisms: Ontology, Agency, and Politics*. Duke University Press, Durham, North Carolina, and London.

Crossland, Zoë. 2009. Of Clues and Signs: The Dead Body and Its Evidential Traces. *American Anthropologist* 111(1):69–80.

——. 2013. Signs of Mission: Material Semiosis and Nineteenth-Century Tswana Architecture. *Signs and Society* 1(1):79–107.

Cummings, Vicki, and Alisdair Whittle. 2004. *Places of Special Virtue: Megaliths in the Neolithic Landscapes of Wales*. Oxbow, Oxford.

de Certeau, Michel. 1984. *The Practice of Everyday Life*. Translated by Steven Rendall. University of California Press, Berkeley and Los Angeles.

DeLanda, M. 2006. *A New Philosophy of Society: Assemblage Theory and Social Complexity*. Continuum, London and New York.

Deleuze, Gilles, and Félix Guattari. 2007 [1980]. Rhizome: Introduction. In *A Thousand Plateaus: Capitalism and Schizophrenia*, pp. 3–28. Continuum, London and New York.

Derrida, Jacques. 1977. Signature, Event, Context. *Glyph* I:172–197.

Descola, Philippe. 1994. *In the Society of Nature: A Native Ecology in Amazonia*. Cambridge University Press, Cambridge.

Foucault, Michel. 1977. *Discipline and Punish: The Birth of the Prison*. Pantheon Books, New York.

Fowler, Chris. 2004. *The Archaeology of Personhood: An Anthropological Approach*. Routledge, London.

Fowles, Severin. 2010. People Without Things. In *An Anthropology of Absence: Materializations of Transcendence and Loss*, edited by Mikkel Bille, Frieda Hastrup, and Tim Flohr Sørensen, pp. 23–41. Springer, New York.

Flannery, Kent V. 1967. Culture History v. Culture Process: A Debate in American Archaeology. *Scientific American* 217(2):119–122.

Gell, Alfred. 1992. The Technology of Enchantment and the Enchantment of Technology. In *Anthropology, Art, and Aesthetics*, edited by J. Coote and A. Shelton, pp. 40–67. Oxford: Clarendon Press.

——. 1998. *Art and Agency: An Anthropological Theory*. Oxford: Clarendon Press.

Gibson, James J. 1979. *The Ecological Approach to Visual Perception*. Erlbaum, Mahwah, New Jersey.

Giddens, Anthony. 1984. *The Constitution of Society: Outline of the Theory of Structuration*. University of California Press, Berkeley and Los Angeles.

Gosden, Chris. 1994. *Social Being and Time*. Blackwell, Oxford.

——. 2005. What Do Objects Want? *Journal of Archaeological Method and Theory* 12(3):193–211.

Haraway, Donna. 1985. Manifesto for Cyborgs: Science, Technology, and Socialist Feminism in the 1980s. *Socialist Review* 80:65–108.

———. 1991. *Simians, Cyborgs, and Women: The Reinvention of Nature*. Routledge, New York.

———. 2003. *The Companion Species Manifesto: Dogs, People, and Significant Otherness*. Prickly Paradigm Press, Chicago.

———. 2008. *When Species Meet*. University of Minnesota Press, Minneapolis.

Harris, Oliver. 2012. (Re)assembling Communities. *Journal of Archaeological Method and Theory*. Published online June 2012, DOI 10.1007/s10816-012-9138-3. Last accessed May 20, 2013.

Hawkes, Christopher F. C. 1954. Archaeological Theory and Method: Some Suggestions from the Old World. *American Anthropologist* 56(2):155–168.

Heidegger, Martin. 1962 [1927]. *Being and Time*. Translated by John Macquarrie and Edward Robinson. Harper and Row, New York.

Henare, Amira, Martin Holbraad, and Sari Wastell, (editors). 2007. *Thinking Through Things: Theorizing Artefacts Ethnographically*. Routledge, London.

Hodder, Ian. 1982. *Symbols in Action: Ethnoarchaeological Studies of Material Culture*. Cambridge University Press, Cambridge.

———. 1986. *Reading the Past: Current Approaches to Interpretation in Archaeology*. Cambridge University Press, Cambridge.

———. 2011. Human-Thing Entanglement: Towards an Integrated Archaeological Perspective. *Journal of the Royal Anthropological Institute* 17(1):154–177.

———. 2012. *Entangled*. Wiley-Blackwell, Oxford.

Horst, Heather and Daniel Miller. 2006. *The Cell Phone: An Anthropology of Communication*. Berg, Oxford and New York.

Horst, Heather, and Daniel Miller, (editors). 2012. *Digital Anthropology*. Berg, Oxford and New York.

Husserl, Edmund. 1960 [1931]. *Cartesian Meditations: An Introduction to Phenomenology*. Translated by D. Cairns. Kluwer, Boston.

Ingold, Tim. 1993. The Temporality of the Landscape. *World Archaeology* 25(2):152–174.

———. 2000. *The Perception of the Environment: Essays on Livelihood, Dwelling, and Skill*. Routledge, New York and London.

———. 2007a. Materials Against Materiality. *Archaeological Dialogues* 14(1):1–38.

———. 2007b. *Lines: A Brief History*. Routledge, London and New York.

———. 2008. When Ant Meets Spider: Social Theory for Arthropods. In *Material Agency: Towards A Non-Anthropocentric Approach*, edited by Carl Knappett and Lambros Malafouris, pp. 209–215. Springer, New York.

———. 2011. *Being Alive: Essays on Movement, Knowledge, and Description*. Routledge, London.

———. 2013. *Making: Anthropology, Archaeology, Art, and Architecture*. Routledge, London.

Jones, Andrew. 2007. *Memory and Material Culture*. Cambridge University Press, Cambridge.

Joyce, Rosemary. 2003. Concrete Memories: Fragments of the Past in the Classic Maya Present (AD 500–1000). In *Archaeologies of Memory*, edited by Ruth M. Van Dyke and Susan E. Alcock, pp. 104–125. Blackwell Publishers, Oxford and Malden, Massachusetts.

———. 2008. Practice In and As Deposition. In *Memory Work: Archaeologies of Material Practices*, edited by Barbara J. Mills and William H. Walker, pp. 25–40. School of Advanced Research Press, Santa Fe.

Keane, Webb. 2003. Semiotics and the Social Analysis of Material Things. *Language and Communication* 23(3/4):409–425.

———. 2005. Signs are not the Garb of Meaning: On the Social Analysis of Material Things. In *Materiality*, edited by Daniel Miller, pp. 182–205. Duke University Press, Durham, North Carolina.

Knappett, Carl, and Lambros Malafouris. 2008. Material and Nonhuman Agency: An Introduction. In *Material Agency: Towards a Non-Anthropocentric Approach*, edited by Carl Knappett and Lambros Malafouris, pp. ix–xix. Springer, New York.

Kopytoff, Igor. 1986. The Cultural Biography of Things: Commodification as Process. In *The Social Life of Things*, edited by Arjun Appadurai, pp. 64–94. Cambridge University Press, Cambridge.

Kubrick, Stanley. 1968. *2001: A Space Odyssey*. Motion picture, Metro-Goldwyn Mayer.

Küchler, Suzanne. 2002. *Malanggan: Art, Memory, and Sacrifice*. Berg, Oxford and New York.

Küchler, Suzanne, and Daniel Miller, (editors). 2005. *Clothing as Material Culture*. Berg, Oxford and New York.

Latour, Bruno. 1993. *We Have Never Been Modern*. Translated by Catherine Porter. Harvard University Press, Cambridge, Massachusetts.

———. 1994. On Technical Mediation—Philosophy, Sociology, Genealogy. *Common Knowledge* 3(2):29–64.

———. 1999. *Pandora's Hope: Essays on The Reality of Science Studies*. Harvard University Press, Cambridge.

———. 2005. *Reassembling the Social: An Introduction to Actor-Network Theory*. Oxford University Press, Oxford.

Law, John, and John Hassard, (editors). 1999. *Actor Network Theory and After*. Sociological Review Monographs. Blackwell, Oxford.

Liebmann, Matthew. 2008. The Innovative Materiality of Revitalization Movements: Lessons from the Pueblo Revolt of 1680. *American Anthropologist* 110(3):360–372.

Marcuse, Herbert. 1982 [1941]. Some Social Implications of Modern Technology. *The Essential Frankfurt School Reader*, edited by Andrew Arato and Eike Gebhardt, pp. 138–162. Continuum, New York.

Merleau-Ponty, Maurice. 1981 [1962]. *Phenomenology of Perception*. Translated by C. Smith. Routledge, London.

Meskell, Lynn. 2004. *Material Biographies: Object Worlds from Ancient Egypt and Beyond*. Berg, Oxford and New York.

———. 2005. Objects in the Mirror Appear Closer Than They Are. In *Materiality*, edited by Daniel Miller, pp. 51–71. Duke University Press, Durham, North Carolina.

Miller, Daniel. 1987. *Material Culture and Mass Consumption*. Blackwell, Oxford.

———. 1998. *A Theory of Shopping*. Polity Press, Cambridge.

———. 2005. Materiality: An Introduction. In *Materiality*, edited by Daniel Miller, pp. 1–50. Duke University Press, Durham, North Carolina.

———. 2008. *The Comfort of Things*. Polity Press, Cambridge.

———. 2010. *Stuff*. Polity Press, Cambridge.

———. 2011. *Tales from Facebook*. Polity Press, Cambridge.

———. 2012. *Consumption and its Consequences*. Polity Press, Cambridge.

Miller, Daniel, (editor). 2001. *Car Cultures*. Berg, Oxford and New York.

Miller, Daniel, and Sophie Woodward. 2011. *Blue Jeans: The Art of the Ordinary*. University of California Press, Berkeley.

Mills, Barbara J., and T. J. Ferguson. 2008. Animate Objects: Shell Trumpets and Ritual Networks in the Greater Southwest. *Journal of Archaeological Method and Theory* 15:338–361.

Mills, Barbara J., and William H. Walker, (editors). 2008. *Memory Work: Archaeologies of Material Practices*. School for Advanced Research Press, Santa Fe.

Munn, Nancy. 1977. The Spatiotemporal Transformation of Gawa Canoes. *Journal de la Societé des Oceanistes* 33(54/55):39–53.

———. 1983. Gawa Kula: Spatiotemporal Control and the Symbolism of Influence. In *The Kula: New Perspectives on Massim Exchange*, edited by Jerry W. Leach and Edmund Leach, pp. 277–308. Cambridge University Press, Cambridge.

Olsen, Bjørnar. 2007. Keeping Things at Arm's Length: A Genealogy of Symmetry. *World Archaeology* 39(4):579–588.

———. 2010. *In Defense Of Things: Archaeology and the Ontology of Objects*. Altamira Press, Lanham, Maryland.

Olsen, Bjørnar, Michael Shanks, Timothy Webmoor, and Christopher Witmore. 2012. *Archaeology: The Discipline of Things*. University of California Press, Berkeley and Los Angeles.

Overton, Nick, and Yannis Hamilakis. 2013. A Manifesto for a Social Zooarchaeology: Swans and Other Beings in the Mesolithic. *Archaeological Dialogues* 20(2):111–136.

Pauketat, Timothy R. 2000. The Tragedy of the Commoners. In *Agency in Archaeology*, edited by Marcia-Anne Dobres and John Robb, pp. 113–129. Routledge, London.

———. 2008. Founders' Cults and the Archaeologies of Wa-kan-da. In *Memory Work*, edited by Barbara J. Mills and William H. Walker, pp. 61–79. School for Advanced Research Press, Santa Fe.

———. 2013. Bundles of/in/as Time. In *Big Histories, Human Lives: Tackling Problems of Scale in Archaeology*, edited by John Robb and Timothy R. Pauketat, pp. 35–56. School of Advanced Research Press, Santa Fe.

Peirce, Charles Sanders. 1992. *The Essential Peirce: Selected Philosophical Writings*, vol. 1:1867–1893. Indiana University Press. Bloomington.

Preda, Alex. 1999. The Turn to Things: Arguments for a Sociological Theory of Things. *Sociological Quarterly* 40(2):347–366.

Preucel, Robert W. 2010. *Archaeological Semiotics*. Blackwell Publishing, Malden, Massachusetts.

Rowlands, Michael. 2005. A Materialist Approach to Materiality. In *Materiality*, edited by Daniel Miller, pp. 72–87. Duke University Press, Durham, North Carolina.

Schiffer, Michael B. 1976. *Behavioral Archaeology*. Academic Press, New York.

———. 1999. *The Material Life of Human Beings: Artifacts, Behavior, and Communication*. Routledge, New York and London.

———. 2007. Some Thoughts on the Archaeological Study of Social Organization. In *Archaeological Anthropology: Perspectives on Method and Theory*, edited by James M. Skibo, Michael Graves, and Miriam Stark, pp. 57–71. University of Arizona Press, Tucson.

Skibo, James M., and Michael B. Schiffer. 2008. *People and Things: A Behavioral Approach to Material Culture*. Springer, New York.

Skibo, James M., William H. Walker, and Axel E. Nielsen, (editors). 1999. *Expanding Archaeology*. University of Utah Press, Salt Lake City.

Shanks, Michael. 2007. Symmetrical Archaeology. *World Archaeology* 39(4):589–596.

Starzmann, Maria Theresia. 2013. Excavating Tempelhof Airfield: Objects of Memory and the Politics of Absence. *Rethinking History* 18(2):211–229.

Steward, Julian. 1955. *Theory of Culture Change*. University of Illinois Press, Urbana.

Strathern, Marilyn. 1988. *The Gender of The Gift: Problems with Women and Problems with Society in Melanesia*. University of California Press, Berkeley.

———. 1991. Partners and Consumers: Making Relations Visible. *New Literary History* 22(3):581–601.

Thomas, Julian. 1996. *Time, Culture, and Identity*. Routledge, London.

———. 2004. *Archaeology and Modernity*. Routledge, London and New York.

———. 2006. From Dwelling to Building. *Journal of Iberian Archaeology* 8:349–359.

Thwaites, Thomas. 2011. *The Toaster Project, or a Heroic Attempt to Build a Simple Electrical Appliance from Scratch*. Princeton Architectural Press, Princeton.

Tilley, Christopher. 1994. *A Phenomenology of Landscape*. Berg, Oxford and Providence.

———. 2004. *The Materiality of Stone: Explorations in Landscape Phenomenology*. Berg, Oxford and Providence.

Trentmann, Frank. 2009. Materiality in the Future of History: Things, Practices, and Politics. *Journal of British Studies* 48:283–307.

Van Dyke, Ruth M. 2007. *The Chaco Experience: Landscape and Ideology at the Center Place*. School for Advanced Research Press, Santa Fe.

———. 2009. Chaco Reloaded: Discursive Social Memory on the Post-Chacoan Landscape. *Journal of Social Archaeology* 9(2):220–248.

Viveiros de Castro, Eduardo. 1998. Cosmological Deixis and Amerindian Perspectivism. *Journal of the Royal Anthropological Institute* 4(3):469–488.

von Uexküll, Jacob. 2010 [1934, 1940]. *A Foray into the Worlds of Animals and Humans with a Theory of Meaning*. Translated by Joseph D. O'Neil. University of Minnesota Press, Minneapolis.

Walker, William H. 1999. Ritual Life Histories and the Afterlives of People and Things. *Journal of the Southwest* 41(3):383–405.

Walker, William H., and Chadwick K. Burt. 2009. New Directions in Late Prehistoric Southwestern New Mexico: Animacy and Archaeology. In *Quince: Papers from the 15th Biennial Jornada Mogollon Conference*, edited by Marc Thompson, pp. 67–72. El Paso Museum of Archaeology, El Paso, Texas.

Walker, William, and Michael B. Schiffer. 2006. The Materiality of Social Power: The Artifact-Acquisition Perspective. *Journal of Archaeological Method and Theory* 13:67–88.

Watts, Christopher, (editor). 2013. *Relational Archaeologies: Humans, Animals, Things*. Routledge, London.

Webmoor, Timothy, and Christopher L. Witmore. 2008. Things Are Us! A Commentary on Human/Things Relations Under the Banner of a "Social" Archaeology. *Norwegian Archaeological Review* 41(1):53–70.

White, Leslie A. 1949. *The Science of Culture*. Parrar Straus, New York.

———. 1959. *The Evolution of Culture*. McGraw Hill, New York.

———. 1975. *The Concept of Cultural Systems*. Columbia University Press, New York.

Winner, Langdon. 1993. Social Constructivism: Opening the Black Box and Finding it Empty. *Science as Culture* 16(3):427–452.

Witmore, Christopher L. 2007. Symmetrical Archaeology: Excerpts of a Manifesto. *World Archaeology* 39(4):546–562.

Zedeño, Maria Nieves. 2008. Bundled Worlds: The Roles and Interactions of Complex Objects on the North American Plains. *Journal of Archaeological Method and Theory* 15:362–378.

——. 2013. Methodological and Analytical Challenges in Relational Archaeologies: A View from the Hunting Ground. In *Relational Archaeologies: Humans, Animals, Things*, edited by Christopher Watts, pp. 117–134. Routledge, London.

CHAPTER TWO

Talk to It

Memory and Material Agency in the Arab Alawite (Nusayri) Community

Şule Can

"Let me tell you about how they make us forget everything that they did to us and how we have been oppressed for centuries," one of my interlocutors started the story. "The tree in Harbiye, I am sure you don't know its story. There is a huge plane tree here, very close to our house. We have asked it to be cut down for at least four years because it was the tree on which Nusayris were executed at the time of [the] Ottoman Empire, and they insist on not doing it. Why? Because they do want to remember how they killed us. We want them to cut it not because we want to forget, we won't, but this shows how they don't care about us and about our history."
—Mehmet, age 50, Harbiye, Hatay

One of the most problematic issues in identity politics, oral history, and memory studies is the history of minorities and the ways in which they are represented. The quotation above is a complaint, a reproach of an Arab Alawite man in a conversation that we had during fieldwork.[1] The tree is not only a representation of political conflict and violence or cruel past actions but also the concrete entity through which people experienced a lot of deaths of their ancestors. The tree is a witness to those deaths and is to some extent responsible for them; but more importantly, its existence means the continuity of oppression. Material culture, in this regard, has become central to being able to remember and forget and to make sense of past experiences. One of the aims of this chapter is to argue how materials matter, in political and religious realms, and how "things" impact people's lives in the Arab Alawite community.

Material culture has tended to be marginalized in contemporary social science research despite the fact that things are a fundamental part of social life (Miller 2005). Recently, there has been a resurgence of interest in material culture on the part of ethnographers and archaeologists, who have made important contributions to building theories of material life (e.g., Gosden 2005; Hodder 2011, 2012; Miller 2008; Mills and Walker 2008; Olsen et al. 2012; Watts 2013). Some basic questions illustrate why it is important to turn to material culture studies. For example, "what makes us human? Are we human because of intentional behavior? Symbolism? Ideology?" (Webmoor and Witmore 2008:55). Further, "what do things do? Does that question attach a power or capacity to objects?" (Gosden 2005:196). There may not be certain and simple answers to these questions, but there is no doubt that our selves and identities are shaped by the way we experience the material world and by what we possess or consume. Moreover, practices or experiences with the material world are not disconnected from it, and engagements with it depend heavily on specific cultural contexts and historical conditions. Therefore, we must think about objects as we contemplate the mutual constitution of people and things (Gosden 2005:197). Humans, collective memory, and oral history are embodied through the material world. When analyzed contextually, namely, in specific cosmologies, the concept of materiality becomes central in terms of considering the ways in which people and objects interact. The way to interpret this interaction does not necessarily mean that people merely attach meanings or give roles to the objects; this interpretation reifies Western dualisms. It is essential to underline the fact that memory, remembering, and materiality and agency need to be understood in their own terms within certain contexts.

Central to understanding materiality is its relationship to agency (Latour 2005; Mills and Walker 2008:14). The concept of agency used in this study is more grounded in Alfred Gell's (1998) *Art and Agency*. A Gellian concept of agency is relational, and human agency is exercised in the material world. One of the essential points in Gell's theory is to understand primary and secondary agents and how they are applied in the examples that he presents. *Secondary agents* for Gell are entities not endowed with will or intention by themselves. Rather, they are essential to the formation, appearance, or manifestation of intentional actions. Secondary agents borrow their agency from an external source, which they mediate and transfer to the *patient* (Gell 1998:34). The index is also another key term in Gellian analysis of secondary agency. He defines it as "an entity from which an observer can draw a causal inference of some kind, or an inference about the intentions or capabilities of another person" (Gell 1998:14). An example of this can be visible smoke, betokening fire. Fire causes smoke, hence smoke is an index of fire. As a pivot of the art *nexus*, the index almost always carries secondary agency, and intentional agency lies outside of it. The artist and the recipient are *primary agents*, but the index is an instrument of social agency in an art nexus (Gell 1998:63). One of the main aims of this study is to explore how materiality and agency are perceived and experienced in a specific context and community that employs Gellian concepts of agency. This chapter also illustrates how we can use Gellian ideas to transcend the structural limitations of the art nexus and engage with the historical and political forces entangled in material agency.

The first part of this study focuses on the construction of cultural memory through materiality and the role of materiality in Arab Alawite religious rituals, belief systems, and community. The complexity of understanding the materiality in the case of Arab Alawite community is that the epistemology of Nusayrism consists of many differences with a Western understanding of the material world. Therefore, materiality in that sense includes particular locations where rituals take place and things that can be considered at the core of the construction of ontological meanings of the belief system, as well as memory. Another part of the complexity stems from the political position of Arab Alawites as an Islamic minority in Turkey, as a result of which cultural memory and materialization of memory become more significant. The second part of the paper explores a Gellian (1998) perspective toward objects as agents.

Nusayri Belief System and Diaspora

The Arab Alawite community is an Islamic community in southern Turkey—the region of Adana and Hatay (Figure 2.1).[2] The Arab Alawite population is around one million in Turkey and around five million in the world. Nusayrism departs from mainstream Sunni Islamic thought in terms of its esoteric character and its inclusion of certain historical events that are not included in the mainstream history of Islam's teachings (Can 2011). Particularly, one historical moment is the key point when Arab Alawites shifted away from the mainstream Muslim community. The first event is Gadir Hum (Eid al Gadir), in which the Prophet Muhammad, just before he died, chose Imam Ali as the caliph. However, Sayyidna (which means our master) Ali was not allowed to be caliph. Sunni Muslims (especially the leader Abu-Bakr) betrayed him, and Abu-Bakr became

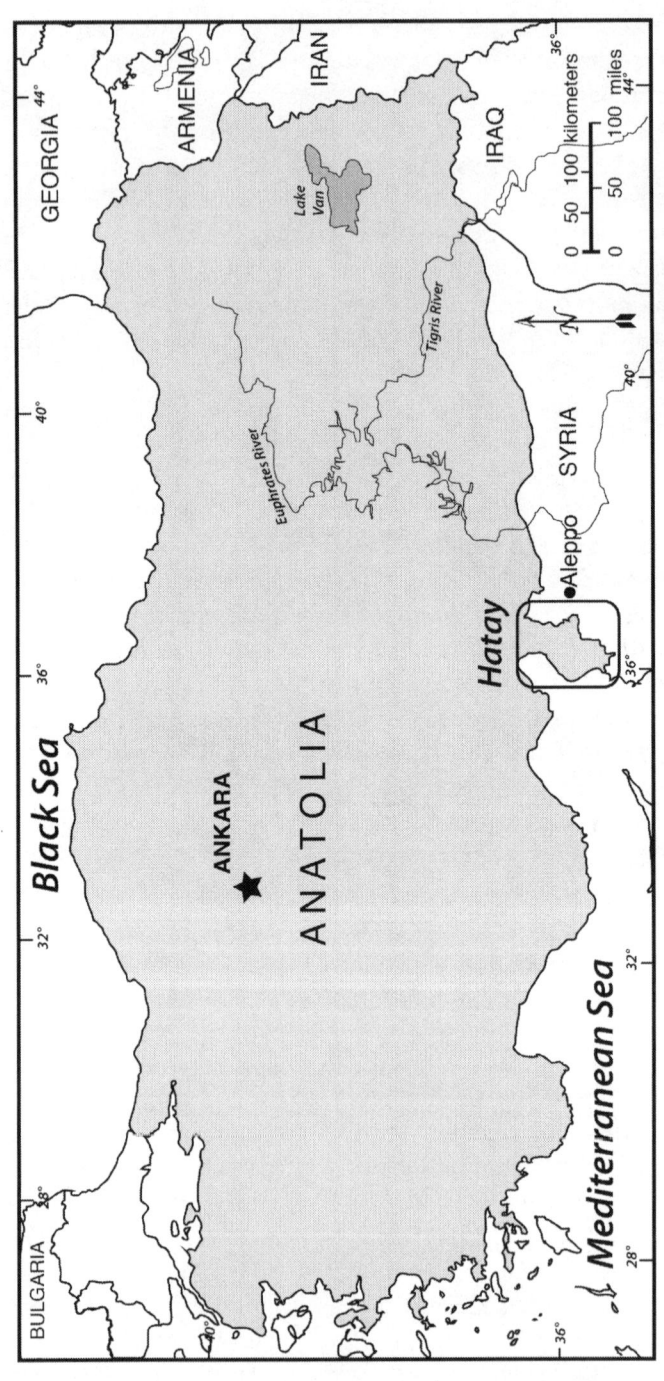

Figure 2.1. Location of the city of Hatay and the Arab Alawite diaspora on the border of Turkey and Syria. Drafted by Anne Hull.

the caliph. This event is core to the Arab Alawite community, as they believe that they are the followers of Imam Ali. And they have never betrayed the orders of the Prophet (who takes his orders from Allah).

The Nusayri belief system is complicated, and it is difficult to make generalizations about all aspects of the religion due to its esoteric character. The Arab Alawite community is based on a secret shared by the men of the community. The heart of the Arab Alawite faith is the belief in One God, "the Essence" (*ma'na*) and Eternal Light, who is accompanied by his "Name" (*ism*) or "Veil" (*hijab*) and his "Gate" ([*bab*] Procházka and Procházka-Eisl 2010:82). In the present Islamic cycle, the Trinity has been incarnated as Ali (*ma'na*), Muhammad (*ism/hijab*), and Salman al-Farisi (*bab*). Hence, it is called "the concept of *Ayn-Mim-Sin*." The twelve imams are regarded as manifestations of the eternal God and are accompanied by a person called the Gate. The eponymous founder of the Alawi religion, Muhammad ibn Nusayr, is regarded as the bab of the eleventh imam, Hasan al Askari. However, Arab Alawites relate themselves to the times before Muhammad ibn Nusayr; therefore the origin of Nusayrism is still controversial even within the Arab Alawite community and among community leaders. This brief summary of the belief system is given in order to understand the *batini* (esoteric) aspect of the religion.

My motivation to investigate materiality, memory, and agency of the sacred places in the Arab Alawite community is derived from my ethno-religious background and experiences as a member of the community. Born and raised in an Arab Alawite town, I am familiar with the sensitivities of the community, misrepresentations or underrepresentations, and ways to think about the abstractions within the belief system. Being a member of the community has also allowed me to be able to conduct long-term observations and have in-depth conversations and interviews with religious leaders, as well as ordinary people. My engagement with the "sacred places" and my indexicality as an anthropologist has brought another layer into the interconnectedness of materiality and memory with ethno-religious identity and agency.

One of the important points to be noted is the current position and religious/political struggles of the Arab Alawite community in Turkey. An assimilation process has led toward homogenization of different ethno-religious identities; this is a determinant factor in the issues of remembering and forgetting in the Arab Alawite community. Historically, different ethnic minorities have been marginalized particularly with control mechanisms that strengthen nation-state ideology, which conceives secularism and modern "Turkish" ethnic identity as its founding principles (Can 2011:40).

Therefore, the only way for the Arab Alawite community to exist in modern hegemonic Turkish discourse and context is to maintain religious practices, sacred places, and native language—namely, Arabic. Furthermore, the esoteric basis of the belief system complicated the continuity of piety and religious practices as informal religious education also declined. But in this case, religious identity and political identity are intertwined. Thus, materialization of memory and perpetuation of agency of sacred materials are also representations of a political struggle and a political agency. Therefore, the next section discusses how sacred places and Arab Alawite rituals challenge forgetting and their roles in materialization of the social memory.

Rethinking Cultural Memory and Materiality

Wertsch highlights Nora's dictum that "we speak so much of memory because there is so little of it left" (Nora 2010 quoted in Wertsch 2004:30), which seems quite apropos when trying to account for today's widespread concern, if not anxiety, about the issue. No matter the motivation, memory has become a major topic of discussion in many disciplines, such as psychology, anthropology, sociology, and history. The term has also come to be used in many different ways; however, its power still resides in its ability to create, form, refashion, and reclaim identity. As Rodriguez and Fortier (2007:9) argue, memory is only one modality of experience. It does not exist in isolation from those other modalities that are oriented to the present and future. The way we remember, what we anticipate, and how we perceive is largely social.

Ideas are abstract, but remembering, according to Jan Assmann, is concrete. Assmann emphasizes that concepts and experiences constitute figures of remembering, which can be categorized as dependent on time, group, and construction of history. Cultural memory is ritualized through monuments, writing, or ceremonies (Assmann 2001:133). The centrality of cultural memory stems from its inevitable role in the construction of collective identity. With regard to the importance of cultural memory, a question that Connerton (1989:70) addresses has evoked a lot of consideration: "how do societies keep their memories, and how do they rediscover them?" This question brings us to the intersection of materiality and cultural memory. Interactions with the material world are important components for construction of memory and social identity. Hence, in this work, cultural memory is used as a concept consisting of narratives, repetition of rituals, or (re)construction of the past and dependence on the community or collectivity. More importantly, it focuses on how the material world gains significance in order to substantiate the narratives and key ideas within the belief system and to provide permanence of religiosity and solidarity within the Arab Alawite community.

Sacred Places, Sacred Bodies

Everydayness of the Arab Alawite community is re-created daily through collectivity and collective actions, such as traditional weddings, visits on the sacred days, or visiting the sacred places as families. Religious elements and practices are a very important part of everyday life in Arab Alawite region and community. The only places—other than houses—to pray and fulfill the religious requirements for Arab Alawites are sanctuaries called *zyaras* (Figure 2.2). There are a few different types of zyaras. For instance, some are dedicated to persons who are not really buried at the place; thus they are called *maqam* instead of zyara. Others include places where sacred events took place in history, tombs of sacred individuals, such as sheikhs (community leaders), or tombs of *wali Allah, ewliya* (saints). Venerated persons for whom zyaras are built can vary, but they

Figure 2.2. A zyara in Antakya. Photo by Şule Can.

can be separated into five categories: prophets; Ahl-al Bayt (the family of the Prophet Muhammad); companions of the Prophet Muhammad; well-known persons of Alawi history; and finally, heroes, fighters, and sheikhs.

For Arab Alawites, the sacred place and its venerated resident are indistinguishable. Women especially use the term zyara for the sanctuary, as well as for its spiritual owner and his deeds, as in "and then the zyara came and did this and that." In their ethnographic study, Procházka and Procházka-Eisl (2010) also state that the sanctuaries do not belong to a particular person or family; therefore zyaras can be seen as community heritage, and the government officially recognizes them as shrines.

> The Arab Alawites often regard the construction of a sanctuary building not so much as the activity of the person who actually takes the initiative but rather as the activity of the Saint himself, who has demanded that he be venerated in a building, not merely at a tree or cenotaph. However, those who decide to erect, rebuild, [or] renovate a sanctuary structure often simply wish to do something good, mostly as the fulfillment of a vow or as an expression of thankfulness towards [the] Saint. (Procházka and Procházka-Eisl 2010:115)

Narrations and myths also play an important role in the sacredness of these places, because Arab Alawites can build zyaras based on some unusual events, which must be understood in terms of batini meanings of the belief system. For instance, a person can have a dream about a saint or a sacred figure or can promise to build a zyara dedicated to a sacred person on the condition that his or her wish is accepted. However, these cases are not ordinary events but rather are miraculous. In fact, what is believed in these cases is that the person is heard by the saint; thus a zyara must be built in return in the name of the saint. It is important to note that there is no direct translation for the names used for the sacred people, so "saint" is used for practical reasons. The sacredness of the saints has enormous significance for the practice and function of the sacred places. The veneration of the saint depends on the belief that a holy person's spirit has a manifestation indicating *tashrifa* (honoring visit). Therefore, the zyaras are built in these places where the spirits manifest themselves. These spirits are believed to be close to God (Allah); they love the Ahl-al Bayt; they have a deep knowledge of God; they have worked miracles; and they follow the example of the Ahl-al Bayt by their righteousness.

In the Arab Alawite belief system, "true faith" and the love of God are the essential practices. The other practices of mainstream Sunni Islam are also required; however it is not enough to be a true believer. True faith is the love of God, Muhammad, and the Ahl-al Bayt and belief in Ali's sanctity. Therefore, visiting sanctuaries is a duty and a spiritual need. The concept underlying a visit to a sanctuary is that of *baraka*, the divine blessing. The *ruh*, or soul of the venerated person, which is believed to be present at the sacred place, is seen as a mediator between God and the individual, and all various rites that pilgrims perform at the sanctuaries are aimed at receiving a bit of the place's baraka. Through the ruh of the saint or prophet, who is also called the Sahib al-maqam, this baraka is directly connected to the sacredness of the place. Thus, one has to actually visit a shrine to get the help of a saint because this is the only way the saint will bestow his baraka. A prayer at home to the saint usually would not suffice (Procházka and Procházka-Eisl 2010:186).

The significant point here is the fact that the only way to get help from saints is to "be" in the sanctuaries or zyaras. This necessity means that there must be a material existence in terms of the place, as well as the bodies that interact with the sanctuaries in a way that makes it unique. Arab Alawites believe that even if one does not pray or speak a word in the zyara, she or he can still perform a ritual just by "being there" because the practice itself is already a part of spirituality and existence of religiosity. Therefore, Arab Alawites who migrate to western or northern Turkey or abroad always feel a loss of access to the sanctuaries. In that sense, the materiality of these places is a key part of practicing the religion or simply continuity of faith. Not only does it change the meanings and dynamics of the city by perceiving the space as a basis of spirituality, but it also creates a bonding between the town and people. As a result, the "community" reconstructs the collectivity, conviviality, and identity with materiality. It establishes more than a sense of belonging, and it materializes "homeland." In this case, homeland is not represented through an identification with a nation or a country at large but with a performance of a minor regional identity as a diasporic space.

Another important point to note is that zyaras are absolutely inalienable and cannot be removed or relocated. Exhumation and relocation of a holy person has always been regarded as a sacrilege. The belief is that people who interfere in the sacred site will be punished, and violators will have to face the consequences. There are many legends concerning what happened when a shrine had to be moved for, say, street or factory construction. In most cases, the saint defended his sanctuary by crippling power shovels or injuring workers (Procházka and Procházka-Eisl 2010:86).

Tilley (2004) emphasizes the use of material metaphors as devices not just for expressing, but also for understanding. An important way material metaphors work is by appealing to the experiential and the nonlinguistic (Boivin 2008:56). What can be recognized from this claim is that zyaras also should not be seen merely as metaphors or simple representations of the saints. Zyaras can be seen in two ways. First, within Nusayri cosmology, the esoteric concepts and meanings, which are mostly difficult to comprehend or understand, become intelligible through our experiences with zyaras and their materiality. Second, transmission of esoteric knowledge and memory is not feasible without zyaras and bodily practices taking place in them. Therefore, in order to explain basic concepts and divine aspects of the religion, zyaras lead the way to make a connection between the material and the immaterial. Experience in the zyara is the sacredness of the saint or prophet and the materiality of the zyara at the same time.

The material existence of zyaras and the engagements with them are also a part of materialization of memory. The community keeps the continuity through the sacred places, rituals, and narrations. Zyaras as places for collective experiences impact social relationships, as well as cultural memory. Paul Connerton presented two ways in which the performative, habitual aspects of commemorative practices produce social memory: through incorporating bodily practices and through inscribing practices, including writing (Connerton 1989:27). These practices should be viewed as working together to illustrate how memory is transmitted and embodied through practice.

Gell describes how Maori meetinghouses can be seen as the collective production of many separate artists and builders, working in separate communities at different times. Although each was striving to produce something distinctive, all are expressions of a common historical trajectory, a common cultural system, and a common ideological and political purpose. We are entitled, therefore, to group them together since they constituted a common pathway for the physical expression of Maoridom as a collective experience (Gell 1998:252).

More importantly, as Gell states for the Maori houses, "the house is a body for the body." To enter a house is to enter a mind, a sensibility. It is to enter the belly of the ancestor and to be overwhelmed by the encompassing ancestral presence (Gell 1998:253). In that sense, Arab Alawite sanctuaries in general are "sacred places for sacred bodies or names" that materialize a total abstract belief system and let people live in the region collectively. Therefore, the power of sacred places cannot be ignored. What they do cannot be argued within the frame of typical understanding of agency or textual representation. Rather, what they do can be seen only in terms of the ways in which their material existences are entangled with existence of Arab Alawites in that specific context or region. Building and protecting zyaras presents materiality as a medium of symbolic construction of the community (Cohen 1985) and more importantly reproduces the Arab Alawite diaspora as sacred landscape.

Initiation Rituals: Collectivity and Memory

The most important rituals for Arab Alawites are initiation rituals through which every man of the community becomes a true member. The initiation of young men into the secrets of religion is carried out by an *amm sayyid* (religious instructor), and usually begins at ages fourteen to sixteen. The initiation lasts for nine months and intentionally parallels the period of normal human gestation. It is clearly meant to indicate that the result of the initiation is a new being possessing the batini knowledge not vouchsafed to non-Alawis (Procházka and Procházka-Eisl 2010:90). The ritual takes place in the young candidate's house. The ritual has three stages, and the third stage is the ninth month in which the candidate goes with his instructor and literally disconnects himself from the outside world. This process is considered as a transition into adulthood, through which the boy achieves a full totality. The secret knowledge he learns is a part of cultural memory and a teaching. Therefore, after the initiation, young men can *exist* in the community and can attend the ritual practices that are carried out in every feast (around twelve feasts a year).

Arab Alawite feasts have two meanings: an outer (zahiri) and an esoteric (batini) or secret meaning, the latter of which is for the Alawis the *real* significance of the feast. In every feast, men conduct the ritual in the houses or certain congregational places, and they are led by sheikhs. The men pray while sitting and facing the sheikhs. There are also a few rites performed during the prayers, such as the mutual handshaking and peace blessing. Rites involve an important bodily and experiential dimension that lies beyond abstract conceptualization, and rites frequently make sense for people at the level of experience (Boivin 2008:85). Boivin argues that it is crucial to see nonlinguistic understanding of the experience or rituals. She invites us to think of the emotional and sensuous aspects of rituals rather than mere symbols.

The sensory and emotional aspects of material experience are not directly comparable to linguistic experience; instead, they demand that we consider various nonlinguistic, supposedly "irrational" aspects of human being. Just as it has been impossible to understand the role of religion within society in the absence of a consideration of experience, so it is impossible to understand the role of [the] material world without giving due consideration to the full variety of ways in which it is experienced. (Boivin 2008:122)

The initiation rituals illustrate how material culture cannot be thought of separately from bodily experiences and social relationships. An important part of the rituals is also a special food called *hrisi* that is shared by as many people as possible. Hrisi is seen as sacred; sharing it means sharing a common belief and "secret" knowledge. However, it is not a linguistic practice. Therefore, the main point is not to pray or use words to receive or give messages. Every person who eats the food becomes part of emotions, intentions, and the ceremony itself. What is significant to understand here is that the very materiality of the food and the practice of sharing it does not represent the sacredness. Rather, it creates or conceptualizes the sacredness and the "feeling" of collectivity and the memory. Similarly, Boivin highlights the strong sensory and emotional experiences often associated with the example of Baktaman ritual experiences, which render them impossible to translate into words. Referencing Barth (1969), she states that "the rite'(s) aspect of doing something . . . clearly predominates for the actors over its saying something" (Boivin 2008:120). The initiation ritual practice of Arab Alawite community is in fact directly related to what Boivin suggests. Body, as inseparable entity from mind or psyche, is expected to be ready to transmit the secret knowledge. However, being ready includes maturity in terms of body, as well as the spirit, which cannot be simply a matter of age or physical development. Space, time, secrecy, mind, or spirit are all embodied in the materiality of ritual practices and the performance of the "body" which cannot be expressed only by means of a linguistic or signification system. The practices that are very mundane in everyday life become a performance through what is said and what is not said in the rituals. Exclusion of direct language use is collective bodily experience on a spiritual level and a way of representing the secret and the pride of sharing it through a particular performance. In fact, what is not said *is* what is performed.

Lastly, the houses where rituals take place are prepared with certain rules. Interestingly, the house as a mundane everyday life space becomes a space for religious activity. The floors of the rooms are covered with consecrated sheets, and incense is burned inside the rooms. The space shifts its meanings as bodies perform there, and the space becomes a part of the sacred activity; although the place is not considered sacred as such.

Ritual Objects: "Being" With/As/For "Them"

The idea of agency is, according to Gell, a culturally prescribed framework for thinking about causation when what happens is supposed to be intended in advance by some person-agent or thing-agent. He emphasizes that he is concerned with the kind of secondary agency that artifacts acquire once they become enmeshed in a texture of social relationships (Gell 1998:17). Hence, his concept of agency is relational, and human agency is exercised in the material world. Bill Sillar (2009) also argues that the material world is not passive, as it provokes and resists human actions simultaneously. However, the degree to which things have secondary or effective agency originates in how people perceive and engage with things through their conscious agency. Analyzing animism in the Andes, he states that animism locates people as participants in the material world and demands that they take responsibility for their relationship with animals, plants, places, things, and people (Sillar 2009:374).

However, it is absolutely not acceptable to talk about any kind of animism while analyzing Arab Alawite rituals or sacred places. The first reason is that Arab Alawite community members are thoroughly Islamic, and they identify themselves as Muslims; animism is not appreciated or a respected idea in Islam. Therefore, it is crucial to understand the complicated ideas of the belief system within its esoteric aspects. The second reason is the main aim of this study, which is to emphasize the importance of understanding each religion, belief system, or cosmology in its own terms and to avoid Western comparisons. Animism refers to a universal category established to make sense of cosmologies that do not distinguish nature/culture or body/mind. Arab Alawite believers accept Islamic doctrine and Nusayrism as true Islam, and within this cosmology, any oversimplified description or categorization, such as "animism," can only represent a reproduction of certain dichotomies.

In this part of the study, I argue that ritual objects or *things* are not only signifiers but constitutive entities that can become the only way of comprehending *presence* and performance of human beings or their intentionality. Materiality is one of the most important elements of the construction and reproduction of memory, particularly in order to transmit the knowledge throughout generations. More importantly, what makes material so prominent is human/nonhuman interaction. Mediating the power of the sacred places, bodily practices, or ritual objects makes them inseparable from materialization of memory. In that sense, Keane addresses some significant questions: "what do material things make possible? What is their futurity? How might they change the person?" (Keane 2005:191). Without doubt, there are very different answers to these questions, and there should be, as material agency or human-nonhuman interaction or entanglement cannot be considered in universal terms. However, the urgency of these questions stems from the quest of looking for new ways to understand material agency without reifying human agency or the subject, whatever it may mean. Therefore, in the next section I tackle the ritual objects and rites in zyaras and the possible distribution of agency between humans and objects.

Rites at the Zyaras

An ordinary visit or typical ritual performed in the sanctuaries consists of short prayers, circumambulations, the burning of *bahur* (incense), and the kissing of the grave(s) and other objects in the shrine. Before entering the holy precinct, the visitor utters the formula *bismillahirrahmanirrahim* (in the name of God). Then, at the beginning or at the end of the visit, bahur is lit. Within zyaras there are usually small places with some coal; bahur is lit there, and a special prayer is recited (Figure 2.3). The bahur has a great amount of power over those who light it and say a specific prayer throughout the ritual. The visitors fan the smoke of the bahur into their faces. The smoke of bahur is believed to protect and relieve the person, and to purify the soul (Procházka and Procházka-Eisl 2010:112). At the entrance to the sanctuary, the shoes are removed. Then two doorposts are kissed and touched with the forehead. This is usually repeated three times. Then, the person circumambulates the grave, which is usually done three times counterclockwise. Since bahur takes the evil away, it is also burned within houses to bring peace home.

An important part of the rite in the zyaras is physical contact and experience. Visitors commonly kiss and touch the doorposts and the grave itself. At some shrines a few special acts of physical touch are performed that are related either to the building itself or to certain objects in the zyara. For instance, *hajrit ish-shifa* (healing stones) are objects that are commonly used to heal some illnesses or pains in the body. These stones are generally cylindrical. Most visitors roll them on their bodies; they may also rub their children with the stones. These stones absorb pain, give strength, and take away *nazar* (evil).

Figure 2.3. The place to light bahur in zyaras.
Photo by Şule Can.

Gell's (1998) framework can help us to understand the role of healing stones within Arab Alawite cosmology. Gell asserts that we cannot discuss index, agent, and patient separately, as they are inextricably interrelated. He also states that unless there is an index, there can be no abduction of agency (Gell 1998:36). Abduction refers to the inference of semiotic meanings from the index. Gell suggests the concept of abduction because the meanings inferred from an index cannot be reduced to a simple iconic or conventional analysis. It needs contextualization to be seen as a natural sign and an index that triggers hedged inferences from which agency can be abducted (Gell 1998:15). Gell's framework emphasizes the efficacy of the object in relation to what human agents infer from it. It is crucial to understand that indexes stand in a variety of active and passive relations to agents. However, this process does not occur without a certain context, which includes communication of meanings and construction of multiple agencies through material objects and human agents as they perceive or imagine them. It is not important whether the healing stone actually heals. Rather, the important point to note is the Gellian presence of multiple agents and patients involving complex relationships (Gell 1998:40). The healing stone is the index as an agent as it heals the person. However, the power of the stone stems from its being in the sanctuary, so what enters into the formula is another agent—the sanctuary itself. But the sanctuary is embedded with the power of the saint—another agent, albeit one that is more esoteric or abstract.

Gell explores non-Western agency on its own terms and in its cosmology. The formula below is used to show how Gell's ideas can help us understand the complex relationship of the objects used by Arab Alawites:

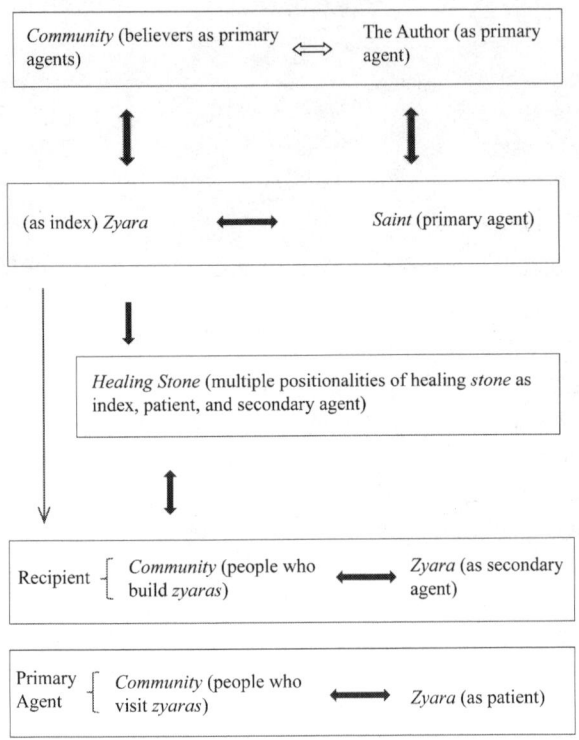

Distributed Agency and Embeddedness of Agent-Patient Relationships

The complexity of human-thing interaction and the ways in which the material world can be seen in constitutive roles problematize agency as a pure human construction and opens up the discussion of *distributed agency*. Things are not primary agents in this formula; however, the interactions take place between people and zyaras, people and healing stones, and finally between zyaras and healing stones. Community members who are believers and zyaras mutually constitute each other. Zyaras are built by community members, although the motivation is basically due to the most important agent, which is the "saint." Zyaras are built and visited by Arab Alawites. Therefore, the complexity in this formula reflects the complexity of human-thing interaction. The materiality of the zyaras is what makes Arab Alawites *be* themselves. However, the reason why zyaras exist is exactly because of the fact that Arab Alawites believe that zyaras are fundamental for them *to be* themselves. Therefore, it is crucial to go beyond a mere engagement with materials and emphasize the dialectical nature of the human-nonhuman entanglement. Another significant point to note here relates to interchangeable roles or types of agency. As Gell puts it, there is no empirical reason why any of our basic "terms" should appear only once in a given formula because in the instances of self-reciprocal agency, the same term necessarily occurs twice, as an agent or a patient. The formula above puts the spotlight on the saint—there cannot be another power comparable to the saint in any way. What Gell does not address in his multiple agency theory is the need to accommodate multiple subject and object positionalities within historical and structural frameworks. The conflict between the Turkish nation-state and ethno-religious minorities is embedded in these constant changes of subject-object relationships, which brings another entanglement into the picture: the materiality of diasporic social space and the materiality of objects in it.

In this nexus, zyaras become powerful secondary agents because of the saints. Healing stones, found in zyara, are material extensions of this agency. Zyaras are built by the people and visited by them; thus, there is a constant mutual constitution of each agent. What Gell claims about art objects is valid for the sacred places or ritual objects; they have no intrinsic nature independent of the relational context (Gell 1998:7).

The indexicality and positionality presented above address the issue of the multiple layers of the construction of meaning. Keane (2003) suggests that signs give rise to new signs, in an unending process of signification. It is important because it can be taken to entail sociability, struggle, historicity, and contingency (Keane 2003:413). He also claims that indexicality and positionality themselves assert nothing. The mediation of signs within a sociological context then might be a way of understanding the social analysis of multiple agents and subjectivity of the materiality. Therefore, in the nexus above, the position that the author takes, as a part of the sociability, and believers who are primary agents brings up the issue of *semiotic ideology* in Keane's view. As an anthropologist who experiences the materiality of sacred places, I become the patient and the agent simultaneously while materializing the experience through writing about it. The social power of the saints, bahur, or zyaras need to be situated within this reflexivity.

Developing such a perspective is not only important but also necessary in order to understand the embedded relationships between the material world, humans, and, in this case, the notion of spiritual existence. Gell proposes such formulations to indicate that there is seamless continuity between modes of action, which involve *performance* and those which are mediated via objects. Every object is a performance in that it motivates the abduction of its *coming-into-being* in the world (Gell 1998:67). When we ask questions related to the origins of objects, the answers are so taken for granted as not to play any part in one's conscious mental life. With objects, as Gell puts it, which are the product of types of agency that we possess generically, the situation is often very different, and we do consciously attend to their origins. This means playing out their origin stories mentally and reconstructing their histories as a sequence of actions performed by another agent, or a multitude of agents, in the instance of collective works, such as cathedrals (Gell 1998:67). The ways in which the objects are historically embodied in our social existence unveil their roles in constituting the world in which we live. This embodiment inserted in our semiotic ideologies not only materializes historical constructions but also *respective futures*, what Keane calls "material's respective vulnerabilities to contingency" (Keane 2003:420).

One of the important examples to understanding the embodied relationship between Arab Alawites and the sacred objects is the practice of "taking things home." Certain things found in many sanctuaries are taken away by the pilgrims because these objects are supposed to contain baraka of the holy place. By taking them home and consuming or wearing them, the pilgrims hope to benefit from the objects' baraka. Many visitors generally take bahur (grains of incense) to use in home rituals or to keep at home because they are believed to be protective. The ritual of the incense can take place at home as well. After burning it in a *mabkhar* (pot), it is taken to each room of the house with certain prayers. This ritual is believed to protect the house from all evil thoughts and intentions and to bring peace at home.

However, nothing can be taken from the sanctuary without leaving something in its place. The objects in sanctuaries are not considered personal property, but the property of the whole community. Furthermore, zyaras need tending, and exchanging objects contributes to the maintenance of these sacred places. The most common way of leaving a material memory of one's visit to a sanctuary is by tying a rag either to one of the window grills of the building or to a tree in or near the shrine. The pilgrims may also bring some stones and leave them on the grave as a sign of their visit. This interaction is a form of accomplishing a spiritual existence through material existence.

When the pilgrims "contact" the saints, this could be interpreted merely as the attachment of meanings to shrines or objects within the shrines. However, this interpretation reifies a categorical distinction that is at odds with Arab Alawite cosmology. Arab Alawites do not think of their existence as apart from all the materials that make them Arab Alawite. Zyaras and objects are part of everyday life and are part of being; they are the materials of future memory, heritage, and resistance to oppression. These places and the objects associated with them are integral to Arab Alawites' ability to say "we exist" as a marginalized community in Turkey. Zyaras, objects, food, and sacred bodies should be understood in an entangled or bundled relationship as bodies who are alive and who maintain cultural memory through rituals, repetitions, sanctuaries, and sacred bodies. The bodies of the saints or sheikhs become meaningful with/as/for material existence of the sacred places and all the sacred places that become meaningful with/as/for the sacred bodies and faith of ordinary people. Thus, zyaras create a type of "boundedness" that would coalesce with the existence of community members. The diasporic significance of zyaras here, then, is the fact that as a marginalized group, Arab Alawites "live together" in a particular region, and without construction of such spiritual attachment, it is not possible to empower the community to resist homogenization and Turkish nationalism.

Conclusion

The reflexive turn in anthropology and the influence of poststructuralism or postmodern critical theorists have given rise to contested claims toward issues of identities and memory. Today, it is possible to talk about how people *experience* these unstable identities and the ways in which they *perform* in their everyday lives. In particular, the focal point is how people experience the material world and how previously inconceivable things may be taken seriously in anthropological discourse. The way people experience or interact with the material world is an inevitable part of their identities and memories. Therefore, remembering is mostly possible through material objects. The practices to remember, namely, memory practices, are seen as a central characteristic of cultural formations.

On the other hand, in developing an anthropological inquiry of materiality and agency, material manifestations of the human condition need to be addressed within social and political systems and actions. Considering conflicts between ethnic groups and politics of identity within nation-state ideologies in Turkey, "remembering" *is* a political action in and for the Arab Alawite community. Existence of Nusayrism is only possible through an imagination of integrity via rituals and material agency in terms of control of the region Arab Alawites live in, as well as in transmitting the memory. Historical construction of "Othering" of "Arabs" in general and Arab Alawites in particular makes cultural memory an agent to reclaim our history, language, and identity. Past experiences that have been kept alive through memory and presence of the sacred places, as well as religious practices, become new representations of resistance to hegemonic Sunni Islam and Turkish nationalist discourse.

Experience is a key aspect to understanding different kinds of human-thing embeddedness because experience is not simply "engaging." A sense of perception is, itself, an act of consciousness from a phenomenological perspective. In this, the reality of the world, as it is naively experienced by us, is suspended—put out of action—not because the reality is in the least doubtful or uncertain, but because it is a product of consciousness, and the aim is to get those activities into view that first of all constitutes this reality. The way we experience the material world reflects the way we form the material world and create a material culture, which forms us in one or another way throughout time. An ethnographic perspective in relation to experiencing the material world hence shows how the theoretical framework suggested by Gell has certain deficits. The first deficit is a lack of emphasis on hierarchy that is inevitably established in culture-specific analyses of materiality. Considering power struggles among ethnic minorities across the world, there cannot be one type of agency that can be endowed with humans and nonhumans to the same degree. Secondly, in Gell's anthropological theory of art, the stress is on aesthetic elements of the objects shown through the art nexus. Gell does not offer a historically informed analysis of primary or secondary agents; rather, he focuses on how artworks act upon their viewers. In the case of Arab Alawites and their sacred places, it is imperative to think about object agency in its direct opposition to hegemonic discourse and "materiality" of discrimination, which cannot be reduced to or seen as a simple performance of an equivalent object-subject agency.

The concept of object agency is problematic and difficult to define. I do not propose an object agency or an oversimplified Cartesian understanding of human-nonhuman agents. However, I do propose a Gellian perspective of secondary agency in order for a better understanding of human-thing engagement or entanglement. The most important reason why I do not present objects as agents is because "human agents" have been dominated or oppressed through power structures and political struggles in contrast to objects, as Van Dyke (this volume) points out. Zyaras are a part of "being" an Arab Alawite, and they "make" identities be as they are to be told and to be proved, which also provides the community with a mutual dependence within the region, as well as continuity of collectivity. With respect to this, it is essential to approach, if not understand, certain engagements with the material world within a specific context considering existential, ontological, and political status of the people or communities.

Embracing Gell's secondary agency is a way of apprehending the materialization of Arab Alawite memory or diaspora, and thus it reveals the limitation of Gell's art-oriented anthropological theory and the ways in which a theory of materiality cannot be detached from its practice. The struggle of Arab Alawites in relation to the sacred places demonstrates that it is essential to shift the attention from merely theorizing the material agency to the production of political minority identities through distributed agency. As Daniel Miller emphasizes, the responsibility of the ethnographer is to document the way different registers of materiality seem to pan out in practice. So the study of material culture often becomes an effective way to understand power, not as some abstraction but as the mode by which certain forms or people become realized, often at the expense of others (Miller 2005:19). In a similar vein, Gell's theory provides us a basis to pursue multiplicity of agency in a material sense. However, incorporating the idea of the materiality of zyaras into a politicized form of agency requires a historically informed analysis through the lens of those who experience them. In other words, historical implications in oppression of minorities in Turkey show that creating a national identity means elimination of any material existence of ethnic and religious minorities. Therefore, the power of the sacred places for Arab Alawites lies in its potential to resist, which Gell missed in his formulation of a non-Western anthropological theory of object agency.

Acknowledgments

This work could not have been possible without the help of many people. First, I cannot thank Dr. Van Dyke enough for her guidance and encouragement in every step of this process. Second, I would like to thank the contributors of the book for their invaluable comments and reviews. The field research in this work was conducted as a part of my MA thesis. I wish to express my gratitude to my MA supervisor Saime Tugrul who guided

me during the field research. Lastly, I also would like to thank my interlocutors in the field and my brother Coskun Can for helping me with the
images.

Notes

1. The terms *Arab Alawite* and *Nusayri* are used interchangeably and refer to the
same community. However, I use the term Arab Alawite due to the current political situation. Community members have expressed concern about the derogatory meanings
attached to the term Nusayri in the Levant; hence, its use has become a controversial
topic. Considering this sensitivity, I prefer to use Arab Alawite as there is a consensus
on the meanings it entails. I refer to the belief system as Nusayrism in this study.

2. The center of the city of Hatay is Antakya; therefore, the city is mostly known as
Antakya (translated as Antioch in English). And it is historically important because it is
a sacred place for Christianity, Judaism, and Islam.

References Cited

Assmann, Jan. 2001. *Eski Yüksek Kültürlerde Yazı, Hatırlama ve Politik Kimlik.* Çevi-
ren A. Tekin. Ayrıntı Yayınları, İstanbul.

Barth, Fredrik. 1969. Introduction. In *Ethnic Groups and Boundaries: The Social Organization of Culture Difference*, edited by Fredrik Barth, pp. 9–38. Universitetsfor-
laget, Oslo.

Boivin, Nicole. 2008. *Material Cultures, Material Minds: The Impact of Things on
Human Thoughts, Society, and Evolution.* Cambridge University Press, Cambridge.

Can, Şule. 2011. Nusayrilik: Sır ve Direniş. Unpublished MA thesis, Department of
Cultural Studies, Istanbul Bilgi University, Istanbul.

Cohen, Anthony P. 1985. *The Symbolic Construction of Community.* Routledge,
London.

Connerton, Paul. 1989. *How Societies Remember.* Cambridge University Press,
Cambridge.

Gell, Alfred. 1998. *Art and Agency: An Anthropological Theory.* Clarendon, Oxford.

Gosden, Chris. 2005. What Do Objects Want? *Journal of Archaeological Method and
Theory* 12(3):193–211.

Hodder, Ian. 2011. Human-thing Entanglement: Towards an Integrated Archaeological Perspective. *The Journal of Royal Anthropological Institute* 17(1):154–177.

———. 2012. Entangled. Wiley-Blackwell, Oxford.

Keane, Webb. 2003. Semiotics and the Social Analysis of Material Things. *Language
and Communication* 23:409–425.

———. 2005. Signs are Not the Garb of Meaning: On the Social Analysis of Material Things. In *Materiality*, edited by Daniel Miller, pp. 182–205. Duke University
Press, Durham and London.

Latour, Bruno. 2005. *Reassembling the Social: An Introduction to Actor-Network Theory*. Oxford University Press, Oxford and New York.

Miller, Daniel. 2005. *Materiality*. Duke University Press, Durham and London.

———. 2008. *The Comfort of Things*. Polity Press, Cambridge.

Mills, Barbara J., and William H. Walker. 2008. *Memory Work: Archaeologies of Material Practices*. School for Advanced Research Press, Santa Fe.

Nora, Pierre. 2010. *Rethinking France: Les Lieux de Memoire Volume 4: Histories and Memories*. University of Chicago Press, Chicago.

Olsen, Bjørnar, Michael Shanks, Timothy Webmoor, and Christopher Witmore. 2012. *Archaeology: The Discipline of Things*. University of California Press, Berkeley and Los Angeles.

Procházka, Stephan, and Gisela Procházka-Eisl. 2010. *The Plain of Saints and Prophets: The Arab-Alawite-Alawi Community of Cilicia (Southern Turkey) and its Sacred Places*. Harrasowitz Verlag, Wiesbaden.

Rodriguez, Jeanette, and Ted Fortier. 2007. *Cultural Memory: Faith, Resistance, Identity*. University of Texas Press, Austin.

Sillar, Bill. 2009. The Social Agency of Things? Animism and Materiality in the Andes. *Cambridge Archaeology Journal* 19 (3):366–377.

Tilley, Christopher. 2004. *A Phenomenology of Landscape: Places, Paths and Monuments*. Berg. Oxford.

Watts, Christopher. 2013. *Relational Archaeologies: Humans/Animals/Things*. Routledge, London and New York.

Webmoor, Timothy, and Christopher L. Witmore. 2008. Things Are Us! A Commentary on Human/Things Relations Under the Banner of a 'Social' Archaeology. *Norwegian Archaeological Review* 41(1):53–70.

Wertsch, James. 2004. *Voices of Collective Remembering*. Cambridge University Press, Cambridge.

CHAPTER THREE

Replicating Things, Replicating Identity

The Movement of Chacoan Ritual Paraphernalia Beyond the Chaco World

Erina Gruner

> Then the seed-priests, the master keepers of possessions . . . gathered
> and fasted, contemplating their sacred objects to divine their meanings.
> It seemed important to them to cut wands from the growth in open
> spaces, to paint them significantly, and add the plumes of the sun-loving
> summer and cloud birds. For they believed that through their incanta-
> tions they could waft the breath of their prayers and their meanings to
> the far-sitting places of the ancients who had first taught them.
> —Origin Story of the Zuni First-Growing-Grass Clan
> (Cushing 1988 [1884]:37)

In today's social science, many are turning away from idealist visions of
society and focusing on redefining the impact of the material. Something
that has emerged from this discourse is an examination of the emergent
agency of objects and assemblages through time—variously referred to
as the *enchainment* or *bundling* of objects and people (Pauketat 2013a,
2013b; Zedeño 2008) or as the *secondary agency* that objects exert on people (Gell 1998). These analyses differ radically from perspectives, such as
actor-network theory (see Latour 2005), which argues that there should
be no a priori analytical distinction between the type of agency exerted by
inanimate objects and that of human actors.

Rather, enchainment and bundling convey the manner in which objects become inextricably associated with other objects, places, and people

and practices in a cumulative historical process—and how the net social impacts of such objects go beyond the intentionality of the person who originally constructed them. When objects occupy a socially laden context, the relationships that surround them can become naturalized to such an extent that the manipulation of one object can impact the entire suite of associated objects, practices, and metaphorical or symbolic qualities in which it is enmeshed; "in other words agency—the capacity to affect social configurations—is mediated by relations in turn contingent on the biographies, genealogies, and histories of larger fields of people, places, and things" (Pauketat 2013a:29).

Thus, although objects are not secondary agents (Gell 1998:17), they nonetheless exert secondary agency through the way that they enchain human relations. These theories, therefore, strike a middle ground between constructions of historical change, which position human actors as mechanistically responding to their material conditions, and social theory, which conflates agency and intentionality.

While discussions of bundling are not exclusively applicable to literal bundles—that is, tied or wrapped medicine kits and fetishes—the theoretical framework is particularly fruitful when applied to religious objects. First, the heightened emotional context of ritual promotes particularly strong preconscious associations between objects and people within this context in the memories of participants. Second, religious objects are often constructed with the specific intent of mediating relations between people and natural forces. The perception that ritual objects have animacy and agency by those who use them heightens their impact on human action. Finally, ritual objects often have particularly complex biographies, and their construction, exchange, and dismantling can enchain numerous actors across broad swaths of time and space.

In the American Southwest, complex bundles, fetishes, staves, and prayer sticks are found at regional centers dating from the tenth century AD to the present. Recent archaeological analyses have emphasized the active role such ritual paraphernalia played as Puebloan and Ancient Puebloan theocracies rose or fell from power (Bernardini 2008; Mills 2004; Webster 2011). In Pueblo Indian societies, where esoteric knowledge constitutes a power resource, power relations can shift significantly when the religious knowledge materialized in paraphernalia is destroyed or appropriated by an opposing faction. However, archaeological analyses rarely address the actual mechanics of how knowledge might be materialized in the attributes of an artifact, which might produce or reproduce a set of social relations through its physicality. If the agency of an object is defined as its ability to act on social relations rather than passively reflect human intentionality, then it is necessary to understand the specific ways in which specific objects "act."

This chapter explores this question using a specific paraphernalia assemblage found at the prehistoric center in the Flagstaff area of northern Arizona and the Chacoan center of Pueblo Bonito in New Mexico (Figure 3.1). A similar assemblage is also used by religious societies at several modern Pueblo communities. I argue that the individual materials of which ritual paraphernalia is composed serve as physical mnemonics anchoring the social, natural, and supernatural relations that the object is intended to effect. The portability of ritual paraphernalia allows for it to serve as the "seed" of complex religious organizations in new communities where specific social roles and identities are replicated in distant areas via the movement of things. Conversely, the destruction of paraphernalia can accomplish the destruction of the collective roles, rights, and identity of the corporate groups that own them.

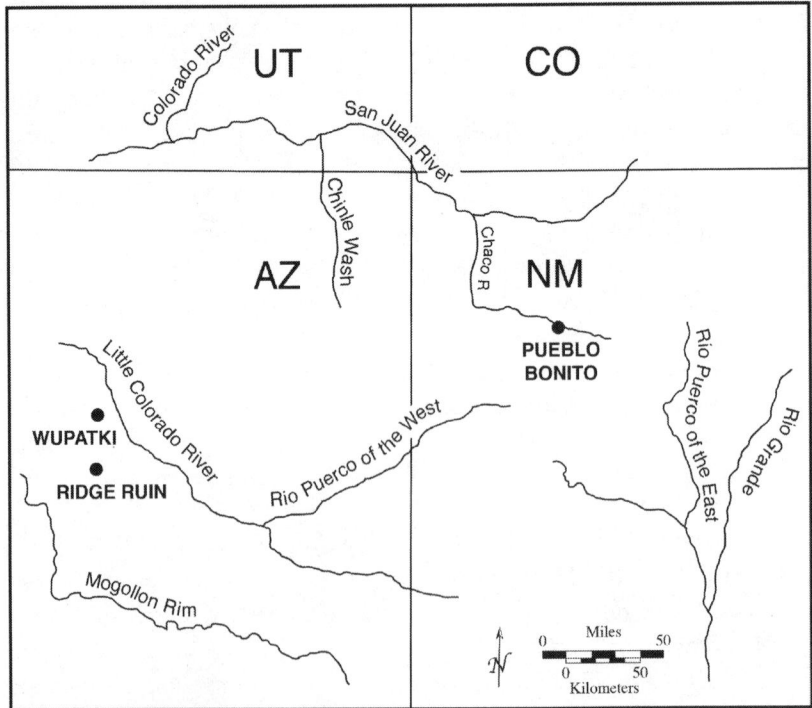

Figure 3.1. Map of the American Southwest showing locations of sites discussed in the text. Drafted by Anne Hull.

Ritual Paraphernalia Assemblages in the Modern Pueblos

The Pueblos are a culturally and linguistically diverse group of communities, and their rituals are likewise diverse; the religious cycle of each community is organized by multiple kin or non-kin-based sodalities (clans or religious societies), some of which are present at multiple Pueblos. Ritual paraphernalia in the Pueblo world is inalienable of individuals, offices, or corporate groups; that is, it cannot be bought or sold in a market exchange, and its life cycle transcends that of its owner. The most powerful sacred objects are explicitly viewed as animate with an agency and intentionality that impacts human relations through supernatural means. Others are believed to channel the agency of their maker according to the properties of the materials that compose them; for example, sticks are adorned with prayer feathers that "carry" supplications to deities and ancestors as a feather is carried by wind (White 1932:127).

Paraphernalia assemblages are often corporately owned and can be curated within a society or clan for generations. However, perishable elements of ritual objects, such as paint, feathers, and various attachments are cyclically renewed by the person who cares for them. Although the paraphernalia assemblage might be used by the society as a whole, the disassembled objects are curated by hereditary leaders or the most highly ranked initiates of a ceremonial society (Parsons 1996 [1939]).

The various objects used by corporate groups form packages that are enchained in ritual practice; thus, the altars of Pueblo religious societies are not permanently assembled furnishings but sand paintings backed by collections of slats, prayer sticks, and other ritual paraphernalia placed in symbolically meaningful arrangements (Figure 3.2). Following the completion of a ritual, the sand is swept away, the curated elements of the altar are disassembled for storage, and impermanent elements are disposed of (Parsons 1996 [1939]:353). Ownership of paraphernalia assemblages therefore entails a certain practical knowledge of how to reassemble objects and assemblages that are perpetually remade.

The repetitive physical practice of constructing and reconstructing complex assemblages serves as a mnemonic that anchors esoteric knowledge, as the process of constructing sacred objects entails the recitation of prayers and oral traditions. This body of knowledge, like the paraphernalia itself, is also inalienable of clans, societies, or offices. The Hopi term *wiimi*, used to refer to an object and the doctrine associated with it, illustrates the inseparability of ritual knowledge and ritual objects; wiimi

Figure 3.2. The ceremonial chamber of a Zuni stick-swallowing society, during their primary ceremony, showing ritual paraphernalia used by various participants arranged into an altar assemblage (Stevenson 1904:plate CVIII).

is "knowledge and practices which are at times embodied in an object" (Leigh Kuwanwisiwma 2007, quoted in Bernardini 2008:494).

There is a generative tendency inherent in the way ritual paraphernalia is materialized within a community. The group or person who curates powerful fetishes or altar assemblages also holds the rights to create other paraphernalia—for example, the myriad of prayer sticks that are constructed for a particular rite and "planted" in sacred places at the ceremony's completion. Certain types of paraphernalia thus serve as the heart of complex object assemblages intended to produce related effects. This generative metaphor is implicit in indigenous terminologies used to describe ceremonial relations. For example, the Zuni use the word *seed* to refer to the powerful *et'tone* (reed-bundle fetishes) of particular societies, and the Tewa have the concept of "re-seeding" extinct societies within a Pueblo community by attracting initiates from other Pueblo groups (Cushing 1988 [1884]; Ortiz 1994:298–299; Stevenson 1904).

Objects thus beget objects. However, perhaps more critically, objects beget categories of people. For example, the names of the social units that historically controlled political action among Pueblo Indians (moieties, societies, and clans) frequently translated to "People of" a totemic object or the natural force its ceremonies relate to—Bow Clan, Wood Society, Turquoise Moiety, and so on.

Perpetual renewal of the ceremonial paraphernalia serves to enchain these groups to their history and territory through specific rules about the social contexts in which the raw materials of paraphernalia might be gathered. For example, individual Hopi clans own the right to gather materials used in their proprietary ceremonies from particular ancestral sites, which are conceived as the "footprints" of their migrations. During pilgrimages to gather materials for ceremonies, Hopis move through the shared history of their clan in a circuit that sometimes covers hundreds of miles, gathering pieces of places to bring back to their home community (Ferguson and Loma'omvaya 2011:146).

Wesley Bernardini argues that Hopi clan migration traditions do not in fact convey the historical movements of a cohesive kin group. Rather, they recount a topogeny of all the places each clan's proprietary ceremony was performed—including areas where social groups other than matrilineal clans likely controlled ceremonies and their paraphernalia. Moreover, because Hopi clan names are derived from a totemic association with the sacred thing they have an affinity for, some traditions recount clan identity changing with the addition of a proprietary ceremony (Bernardini 2008). In essence, then, what these traditions convey is the history of the movement of ceremonies—a material and ideological package—between communities. The relation between clans or societies and their parapher-nalia assemblages is thus circular; the objects produced by the collective produce the collective. From an anthropological perspective, replicating packages of ritual paraphernalia thus provides a continuity of identity and rights as corporate groups migrated from settlement to settlement or new groups rose to prominence.

The replication of paraphernalia assemblages also reproduces the rights that are attached to religious responsibilities: the right to gather powerful materials, to participate in governing councils, to mobilize a labor force for constructing ceremonial structures, and to use prime agricultural land. Achieving rightful possession of powerful ceremonial paraphernalia there-fore impacted relationships between individuals, corporate groups, and villages. Particularly in the Western Pueblo groups—where the core para-phernalia for large ceremonial societies is often owned by the core lineage of a single clan—the movement of a single family between villages could shift the balance of power.

Paraphernalia Assemblages in the Twelfth Century: The Burial of the Magician

When archaeologists discuss the interrelation between ritual objects and traditional ritual knowledge in the prehistoric Southwest, they inevitably reference the "Magician's" burial from the twelfth-century Sinagua site of Ridge Ruin ([Figure 3.3] see Bernardini 2008; Ferguson and Loma'omvaya 2011; Mills 2004). This burial contained a mature male in ceremonial regalia along with some three hundred objects that included numerous turquoise, argillite, shell, and jet ornaments, disassembled ritual parapher-nalia including a reed bundle, painted sticks, coated and painted baskets, inlaid awls, and raw materials, such as pigments, shell, lacquer, and fur, which might have been used to construct other ritual paraphernalia (Mc-Gregor 1943). The burial is one of the few circumstances where a sizable assemblage of (relatively) intact ritual paraphernalia has been found with the religious specialist who used or curated it.

This "Magician"—so named because of his many "magical" objects—was interred in a multilayered, pole-roofed vault in the floor of a kiva. Excavators found other paraphernalia (effigy bows and arrow bundles) scattered across the roof of the grave, and two macaws in cists in the floor nearby. Some of the things buried with or near the deceased—including

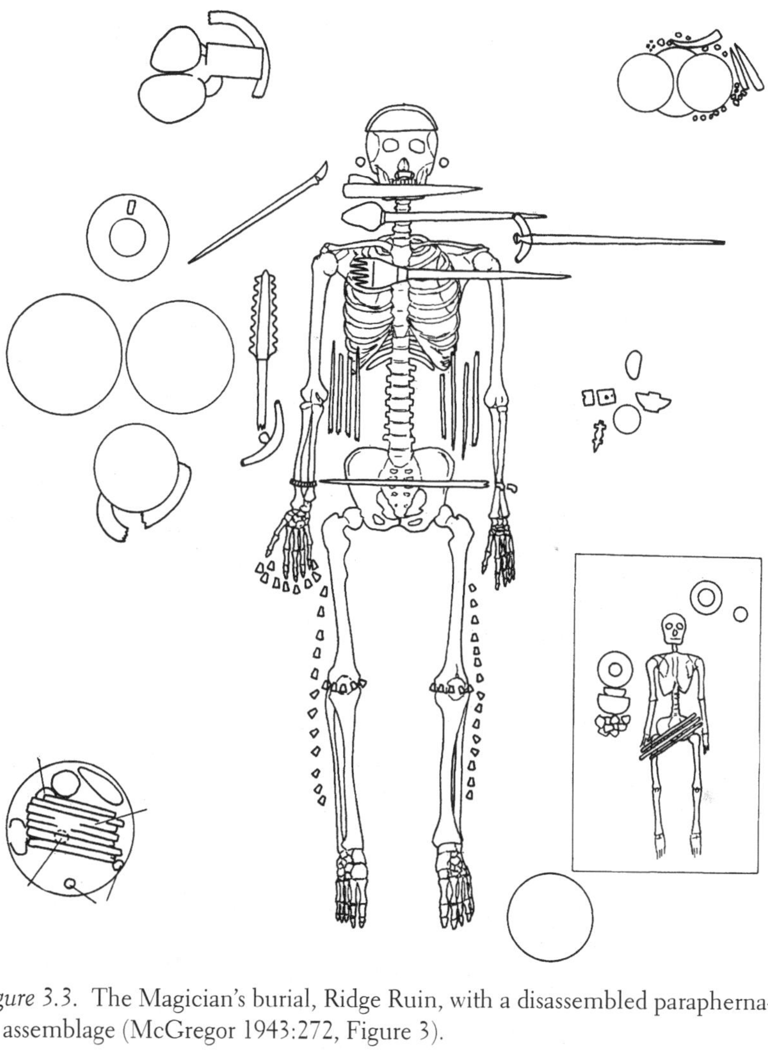

Figure 3.3. The Magician's burial, Ridge Ruin, with a disassembled paraphernalia assemblage (McGregor 1943:272, Figure 3).

macaws and *Strombus* shell—were imported from Mesoamerica and the
Gulf of California and are found at only a small number of prehistoric
centers in the American Southwest (McGregor 1943; Mills and Ferguson
2008; Vokes and Gregory 2007).

This burial assemblage may have remained an enigma had the field
crew that uncovered it not included two Hopi men from Oraibi and Mi-
shungnovi. These men immediately concentrated on the "root" of the assemblage: a series of long pointed, decorated staves known as swallowing

sticks, which were used on the Hopi mesas until the early twentieth century. The purpose of this rite—which was performed by pushing the sticks
down the initiate's throat in an act analogous to sword swallowing—was
war medicine, wherein a person or group would be "strengthened."

The form of the swallowing sticks indicated to Hopi people a series
of relationships between the objects from the burial and between those
objects and particular social actors. For example, a Hopi consultant from
Shungopavi, when shown some of these sticks, accurately described other
objects from the burial including a club-like stick with serrated edges, a
double-horned stick, and a warrior's cap, a sort of peaked skullcap. He
additionally predicted the location of the burial within a twenty-mile ra-
dius based on the fact that the clan that had curated that paraphernalia
at Shungopavi still held the right to gather eagles in the area, which was
presumably their ancestral territory (McGregor 1943:296).

All consultants indicated specific clans were associated with these materials in their villages, although they acknowledged that other clans had
held the ceremony and associated roles in the past. They noted that the
man in the grave was probably a *kaletaka*—warrior or war leader—because
he would typically lead this ceremony (McGregor 1943:295–296). In a
more recent interpretation, Hopi cultural advisors identified additional
raw materials as components used by curers, including mountain lion
teeth and claws, two kinds of shells, and several minerals. They also recognized numerous artifacts as pieces of ritual paraphernalia and costumes,
including a bow guard (decorated wristlet), awls, baskets, fringed leggings,
arrows, and altar pieces (Ferguson and Loma'omvaya 2011:173).

Hopi consultants, in the 1940s as today, have consistently agreed that
the materials represented the holdings of one or more ceremonial societies that were buried with a society leader when he died (see Ferguson and
Loma'omvaya 2011:172–173; Hohmann 1982:55; McGregor 1943:295).
However, such an event occurs only rarely, as destroying such materials effectively destroys the collective identity, rights, and responsibilities that are
anchored in them—destroying altar assemblages can thus literally signal
the death of a clan or society (Whiteley 1998:125–162).

Although discussion of the Magician's burial artifacts typically focus on
Hopi ethnographic analogies, ritual objects like those from the burial are
also used by ritual societies at other Pueblos where they anchor similar
sets of social relations within highly divergent social structures (O'Hara
2008; Gruner 2012). Shared features of stick-swallowing societies at different Pueblos include the intended effects of the rites that utilize these
objects (curing, strengthening, witch hunting, and weather control), the

deities associated with them (the War Twins, Mountain Lion and Bear, and sometimes Knife-Wing Bird), and a shared set of skills. In addition to stick swallowing, many of these societies consume live coals, and fire brands "transform" into bears or mountain lions and "suck" objects out of people's bodies during cures (see Gruner 2012:81–104). Parallels of course can be attributed to historical interaction between allied ceremonial societies at different Pueblos; however, they are also produced by the systems of knowledge and practices embedded in the objects themselves.

On the most obvious level, decorated objects like swallowing sticks encode information about their use through symbolic indexes. For example, the color scheme of swallowing sticks in the Magician's burial probably reflects a pan-Puebloan symbolic complex associating black, white, red, yellow, and blue-green with particular cardinal directions (Parsons 1996 [1939]:99), a significance which is also physically enacted in the cardinally oriented choreography of present-day stick-swallowing rites (see Stevenson 1904:444–515; Titiev 1968 [1944]:157; White 1932:115, 1962:152). Such symbolic aspects of ritual artifacts should not be downplayed simply because our goal is to address their materiality; after all, it is the investment of symbolic attributes in physical materials that might be possessed, destroyed, or manipulated, which makes ritual paraphernalia such a potent force in social relations.

However, more subtly, identification of initiates with the materials they use produces a body praxis and social identity that transcends linguistic boundaries. For example, a Zia informant states that "medicine men eat the fire to get the power of fire; fire is *maiyanyi* or *kopictaiya*" ([i.e., animate] White 1962:153). Likewise, warriors and curers "transform" into bears or lions by donning the claws, teeth, paws, or skins of these animals. This transfers predatory properties to the wearer and produces animalistic behavior, such as growling, crawling, and biting. The premise of using a swallowing stick is likely similar. The complex recipes by which such swallowing sticks are constructed—the blade coated with sacred minerals and the fat of power animals, the handles tipped with power objects, such as prayer sticks, trees from sacred mountains, masks, projectile points, and the plumes of sacred birds (Stevenson 1904; Titiev 1968 [1944]; White 1932, 1962)—bundle multiple animate forces into a single object, which is then "consumed" by the initiate as he swallows it.

Pueblo Indian religion conceives of the transformation of people via sacred things as inherent to those things, which turns the sticks themselves into volatile objects. For example, Parsons notes that two little boys who accidentally touched the swallowing sticks of the Zuni Wood society were initiated for their own health, and sick people who were cured by a stick-swallowing society frequently later joined it (Parsons 1996 [1939]:113; Stevenson 1904:448). In addition, some members joined because their clan owned important paraphernalia and therefore must be represented in particular rites (Stevenson 1904:449). Stevenson notes that this decision was often a reluctant one despite the prestige associated with society membership, as stick swallowing and fire eating were physically demanding and occasionally lethal (Stevenson 1904:449).

During initiation, each person was compelled to replicate his/her "bundles" of core paraphernalia, as he/she was indoctrinated into the system of knowledge associated with it. This system thus produces the peculiar situation of objects reproducing themselves—in part through the actions of people who never wanted them in the first place. Such a system has a viral potential should paraphernalia be introduced into a community—a potential that was first realized in the Sinagua area when the Magician was buried at Ridge Ruin.

Replicating Materials, Replicating Relations

Although the Magician's burial is an anomaly among Sinagua burials, the burial of the Magician was not entirely without precedent in the prehistoric Southwest. In fact, the Magician's burial closely resembles the burial of two mature men in Room 33 at the Chacoan great house of Pueblo Bonito, some three hundred miles away, in a different culture area, and over one hundred years earlier. Pueblo Bonito is presumed to have been the preeminent center of the Chaco system, a network of monumental centers (great houses) that were constructed throughout the San Juan Basin from approximately AD 900 to 1150. These structures were likely occupied by a small group of ritual specialists but used by the community as a whole (Neitzel 2003a).

Like the Magician, the individuals buried in the floor of Pueblo Bonito's Room 33 were interred in ceremonial regalia in a wood-roofed, multilayered vault grave with numerous ornaments, minerals, and ritual objects. Some of these objects were very similar to offerings from the Magician's burial. For example, both burials contained a reed-bundle container, effigy arrows, a turquoise-encrusted mosaic wristlet, turquoise inlay pendants in the shapes of birds and sandals, turquoise ear bobs, *Haliotis* (abalone) shell dippers, a rare *Strombus* shell, and quantities of raw malachite (McGregor 1943; Pepper 1906, 1996 [1920]).

Moreover, the layout of objects within the Bonito vault also resembles the burial of the Magician: objects clustered around the four corners of the grave, across the chest, pelvic area, and to the right of each body (Pepper 1906). Significantly, the skeletal history of Burial 14 and the Magician suggests that they share a similar life history. Both were unusually robust, mature men who showed traumas typically sustained during combat: in the case of the Magician, a defensive wound on his right forearm; and in the case of Burial 14, fatal blows to his skull and chops across his left knee (Akins 2001:172; McGregor 1943:294). In other words, comparison of Burial 14 with the Magician's burial indicates not only the replication of objects but also the replication of the social role of the warrior, the same role held by ethnohistoric Pueblo stick swallowers. While we must assume that the enactment—and moreover the social contexts—of this ritual likely changed radically over the nine centuries between Chaco and the ethnohistoric present, the association between specific ritual objects and a specific type of personal power nonetheless constitutes a core of shared identity with diverse descendent groups.

The sudden death of the man in Burial 14 during combat suggests that Hopi interpretations of the Magician's burial assemblage might also apply here; the unusual quantity of inalienable ritual objects buried with him may have been due to the fact that he had not yet trained an heir in their use and histories—a situation that would have required them to be removed from circulation. Nonetheless, this burial—which occurred fairly early in Pueblo Bonito's occupation sequence—came to be a focus of collective identity in later phases as indicated by the fact that numerous other ritual objects and the bodies of several genetically related persons were later interred in the room above his grave (Akins 2001, 2003; Pepper 1906, 1996 [1920]; Schillaci 2003).

Although many types of paraphernalia found with the Magician's burial are not found in Burial 14 proper, they are found in a discrete section of Pueblo Bonito associated with this kin group. The architecture of Pueblo Bonito is bilaterally symmetrical throughout most of its construction phases and ground-floor access between room blocks was split along an axis running northwest-southeast (Figure 3.4). Within these two sections were two spatially discrete clusters of burials within the western and northern portions of the structure that seem to belong to two biologically distinctive groups (Akins 2001, 2003; Schillaci 2003). Burial 14 was the most elaborately prepared body within the northern burial group, which in general contained a higher volume of paraphernalia and prestige trade goods (Akins 2003; Neitzel 2003b).

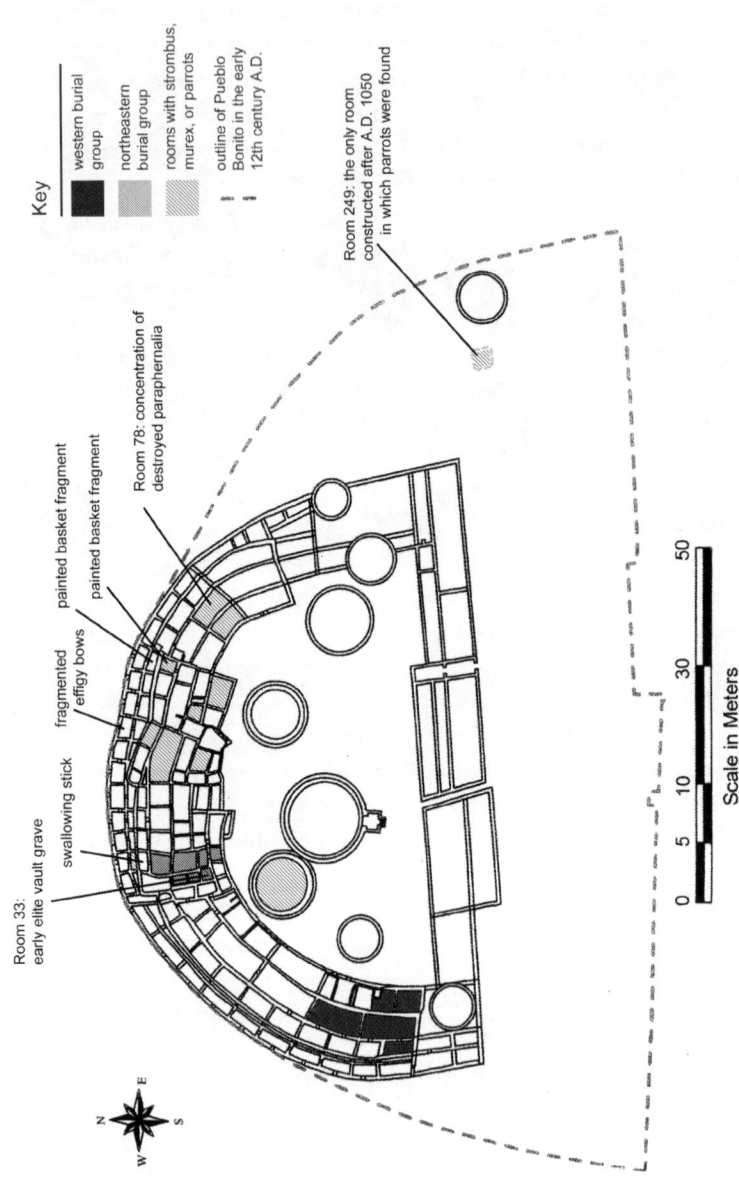

Figure 3.4. Plan of Pueblo Bonito showing the association of the western and northeastern halves of the structure with discrete burial groups during the Classic Bonito phase (AD 1050–1100) and the concentration of types of prestige trade goods and ritual paraphernalia found in the Magician's burial in the northeastern half of the structure. Drafted by Erina Gruner based on Stein et al. (2003, Figure 4.15).

Some archaeologists have proposed that the bilaterally divided architecture of Pueblo Bonito was used in the activities of two different elite kin groups; the western half corresponded to the western burial group, and the eastern half corresponded to the northern burial group (Heitman and Plog 2005). Within Pueblo Bonito, ritual objects similar to the most distinctive artifact types found in the Magician's burial—effigy bows, swallowing sticks, and coated and painted baskets—were exclusively found in the areas associated with the northern burial group prior to AD 1050 (Gruner 2012:134–149). This distribution suggests the possibility that the ceremonies practiced by the Magician were associated with one kin group within Pueblo Bonito, but not the other.

Of course, intact sacred objects are retired only rarely, so the grand total of these objects at Pueblo Bonito amounts to four fragmented effigy bows, two fragments of coated and painted basketry, and one hoof-headed swallowing stick. But when one considers artifacts and materials from the Magician's burial, which are not ritual paraphernalia but which may have become historically enchained with paraphernalia assemblages, the pattern becomes clear. For example, as mentioned previously, one of the social rights that is often attached to ownership of paraphernalia is the right to organize pilgrimages to gather ritually significant plants, animals, and minerals. Within Pueblo Bonito, the distribution of rare shell types found in the Magician's burial—for instance *Strombus* and *Murex* shells and unworked bivalve and *Haliotis* shells—is exclusively confined to courtyard kivas and interior rooms in the same northeastern section of the structure where the ritual paraphernalia was deposited. Likewise, all of the parrots and macaws found within Pueblo Bonito were in these northeastern room blocks, as were the overwhelming majority of jet artifacts and raw jet pieces (Gruner 2012:139–149). This distribution suggests that regardless of whether or not multiple groups were using these materials, the kin group in eastern Pueblo Bonito was probably the one procuring them.

At AD 1050, Chaco Canyon was probably the only location in which this complex package of artifacts, materials, and production technologies co-occurred as an integrated suite. In fact, Mills (2004) suggests that Chaco during this era was likely the first time and place where the prestige of groups analogous to modern Pueblo clans or religious societies became anchored in packages of inalienable ritual paraphernalia.

This is not to say that the two kin groups resident in Pueblo Bonito were the only people using this ritual paraphernalia; by contrast, it is clear that some of Pueblo Bonito's ritual objects were "votive" productions made in widely dispersed areas of the Chaco system but only used and stored by kin groups at Chaco Canyon great houses (Toll 2006). As people gathered in Chaco Canyon for periodic rituals, they brought with them a steady flow of resources and the labor pool used in constructing Chaco's monumental architecture (Toll 2006). The enchainment of particular social groups to particular types of ritual objects and these objects to particular places thus constituted a significant force in Chaco's prestige economy.

Replicating Paraphernalia Assemblages

Archaeologists usually date the demise of the Chaco system to about AD 1140, the point when monumental construction ceased at Chaco Canyon. However, from an artifact-oriented perspective, the beginning of the end occurred in the early twelfth century as Aztec West, a large great house along the northern periphery of the Chaco system, accrued a sizable assemblage of ritual paraphernalia and increasing political independence (Reed 2008; Webster 2011). Conversely, excavations at "McElmo" great houses that were constructed at Chaco during this period are largely devoid of artifacts and production debris; that is, the construction was largely about display and may have been intended to bolster confidence in a failing regime (Van Dyke 2004, 2009).

During this period, people in two centers far beyond the Chaco periphery—Wupatki Pueblo and Ridge Ruin—began to use very similar paraphernalia assemblages (Gruner 2012). This replication was merely one aspect of a general intensification of communal ritual in the Flagstaff area during the twelfth century—a trend which was likely inspired by the combined uncertainties of rapid climate change, increasing population density, and a dramatic volcanic eruption (Elson et al. 2002; Lekson 2008:155–158; Reid and Whittlesey 1997). The nature of the connection between these centers and the Chaco system remains unclear, although technological style of Wupatki's wall construction suggests that some of its founders may have participated in Chaco's programs of monumental construction (Gruner 2012:54–58). However, the distance between these centers and Chaco is formidable, and there is no indication that there was regular interaction or substantial exchange of populations between these areas.

Nonetheless, changes in trade patterns and social hierarchies during this period suggest that it was not only ritual paraphernalia that was imported into the Flagstaff area but also some of the relations associated with them. As swallowing sticks, effigy bows, painted baskets, and other Chacoan artifact types appeared in the Flagstaff area, so did a new prestige technology (turquoise mosaic ornamentation) and numerous exotic items from Mesoamerica and the Gulf of California ([parrots, macaws, and a small number of *Strombus* and *Murex* shells] Gruner 2012:105–123).

The deposition of parrots and macaws in elite burials—including child burials—at Wupatki and Ridge Ruin suggests a system of kin-based control similar to Chaco's (McGregor 1941, 1943; Museum of Northern Arizona site files NA405, NA1785). The labor and materials invested in burials from within these centers present some of the clearest evidence of a pre-historic theocratic elite in the prehistoric Flagstaff region. The apparent influence (and affluence) of these groups suggests that for a short period of time, the ceremonial societies "planted" at these two centers flourished. In this new social milieu, Chaco-style ritual paraphernalia would inevitably have also become layered with new relations and associations—bundled into synthetic rituals enacted in tandem with other ceremonial societies and the mediation of local politics and resources.

The Problems with Materials and Identity

When a social identity is inextricably entangled in things, which pro-duce relations between people, places, roles, and rights merely by their presence within the community, what happens when that social identity becomes problematic? It seems inevitable that such objects would eventu-ally become a focus of conflict; when corporate control over particular re-sources and political offices is predicated on supernatural control over the natural and social order, the society becomes uniquely vulnerable in times of hardship (Levy 1992). The dangerous aspect of ritual power is that the advanced initiate is credited with controlling the outcome of events when harm comes to others within the community. Powerful ceremonialists thus run the risk of being accused of witchcraft should public opinion turn against them. The consequences of such accusations at various Pueblos ranged from public humiliation to torture and execution (Parsons 1996 [1939]:62–67).

There is every indication that the ceremonial roles associated with stick swallowing were regarded as immensely powerful in antiquity and in recent history, and this power was potentially dangerous. For example, Stevenson observes that accused "witches" were often those who were conspicuous in public ceremonies. And in one case, "a man belonging to the . . . [stick] swallowers, which is one of the most important in Zuni, was regarded by the majority of the people as a wizard" (Stevenson 1904:396). Likewise, Hopi consultants repeatedly noted when discussing the Magician's burial that the stick swallowers were regarded as a "witchcraft group" (McGregor n.d.), a situation that may have figured in the widespread abandonment of this practice at Hopi villages around the turn of the last century.

Corporately owned ritual paraphernalia, as a locus of shared knowledge and identity, featured centrally in negotiating such conflicts. For example, the Hopi village of Oraibi split up in the early twentieth century into a ruling faction that advocated a pacificatory stance toward the U.S. government's interventions and an opposing faction that advocated open resistance. During this conflict, the political maneuvers of friendly and hostile leaders materialized through manipulation of ritual objects. As a result, a newly Christianized member of the Bow Clan put his clan's Aa'alt society altar "on trial for murder" and publicly destroyed it (Suderman n.d. quoted in Whiteley 1998:140). As a critical piece in the initiation of new members in a powerful religious society, the altar, by its very existence, held the capacity to reformulate identities and local politics in ways that were beyond its owner's control. Burning the altar not only destroyed the heart of the Aa'alt society, but it also constituted "collective supernatural suicide/deicide on the behalf of his clan—simultaneously abolishing its supernaturally instrumental power and its capacity for sociopolitical legitimation" (Whiteley 1998:144).

Conversely, another Hopi religious leader, the Oraibi Bear Clan headman, decided to be buried with his clan paraphernalia thus depriving his heirs of the religious and political roles associated with its use (Joann Kealiinohomoku 1979 quoted in Hohmann 1982:55). However, in the Hopi conception he did not destroy the ceremony itself because initiates are believed to continue their ritual practice in the afterlife (Ferguson et al. 2001:13). Burying object assemblages or entire ritual costumes with the dead rather than destroying them suggests the desire to preserve the relations enchained in this bundle even while preventing the social dynamic associated with them from being replicated by the next generation.

Disassembling Relations

During the mid-twelfth century, something occurred at Chaco that undermined the foundations of its political authority and prompted the cessation of monumental construction in the canyon. Although most archaeologists suggest that the ecological impetus for this collapse was prolonged drought in the San Juan Basin, the character of material interventions throughout the Chaco world suggest that social and natural stresses prompted a widespread crisis of faith. The most archaeologically visible of these interventions is the burning and filling of ceremonial architecture—events that in some cases can be concretely dated to the mid-twelfth century (Brown et al. 2008:240; Windes 2003:30).

The excavation of a small number of Chacoan great houses suggests that paraphernalia assemblages were also targeted for destruction, although dating this destruction is problematic. For example, excavations in Pueblo Bonito's Room 38—a probable ceremonial chamber in the northeastern area of the structure—contained the burned and fragmented remains of ceremonial pipes, staves, shell trumpets, and effigies along with elite ornaments that included turquoise-inlaid shell and jet and a number of tools by which similar objects could have been produced, such as rasps, hammer stones, grinding stones, and punches. Massed in the center of Room 38 were the bodies of twelve macaws, which had evidently been penned in the room and left to die when it was abandoned. Thus, in one fell swoop, the inhabitants of Pueblo Bonito's eastern portion destroyed a ceremonial chamber, their sacred objects, the means by which such objects could be reproduced in the future, and one of the most striking privileges associated with those objects—the right to raise macaws.

A similar drama is indicated by fragmented paraphernalia assemblages found in Room 93 of Chetro Ketl Great House (Vivian et al. 1978). This room—like Room 78—contained some of the same objects retired in the Magician's burial. As in Pueblo Bonito, these artifacts were left scattered in situ, indicating that this event signaled the abandonment of that portion of the building.

The thoroughness with which powerful ritual objects were disarticulated, burned, and broken at Chaco Canyon suggests the desire to neutralize a potent force by dispersing its constituent elements. One envisions a situation like the burning of the Aa'alt altar assemblage in which the spiritual force animating ritual objects became "corrupt," a circumstance that would have required the total breakdown of the objects themselves and of the knowledge, practices, and social relations embodied in them.

Meanwhile in the Flagstaff area, less drastic transitions were being negotiated. Around the middle of the twelfth century, several rooms within Ridge Ruin and Wupatki Pueblo were burned—in both cases the rooms dated to the initial founding of the center and were built using a distinctive style of masonry associated with Chacoan architecture (Downum 2004; McGregor 1941). However, rather than violently destroying ritual paraphernalia, these communities apparently waited to retire ritual roles with the death of initiated leaders. Excavators with the Museum of Northern Arizona discovered a burial within Wupatki Pueblo matching the Hopi description of "the last survivor of the Parrot clan at Wupatki . . . buried there with parrots at his side" (Stanislawski 1963:62).

Likewise, at Ridge Ruin the burial of the Magician, two macaws, a sizable assemblage of ritual paraphernalia, and the kiva in which they were buried occurred approximately twenty years after the fall of Chaco. The care with which these objects were arranged to preserve spatial relationships to each other and the body of the initiate who used them suggests that obliterating a dangerous spiritual influence was not the goal of the mourners who buried them. Rather, this seems to have been an instance like the burial of the Oraibi Bear Clan headman, a leader who chose not to initiate an heir because he felt that his ceremony was no longer valuable to his community. Sacred objects were therefore retired with the last person who fully comprehended the knowledge that they embodied.

Conclusion

The preceding discussion has attempted to convey the complex ways in which ritual objects act on each other and on the groups that use them. Embodying ritual knowledge and authority in tangible objects that could be transferred, replicated, or destroyed imbues these objects with a secondary agency—with the potential to impact human relations in ways that exceed the intentionality of the ritual specialists who first constructed them. Ritual paraphernalia structures diverse societies in regular ways by prompting specific types of interactions between society initiates, dispersed natural environments, and the communities in which they live. The introduction of paraphernalia assemblages into Pueblo communities can thus act as a transformative force that reconfigures local identities and histories and shifts local power dynamics.

The agency of these objects also holds the potential for their eventual destruction, since the perceived supernatural powers of ritual objects and the tangible rights they uphold prompt conflict in times of social stress. However, the actual impacts of ritual objects on human societies remain invisible until relations between the objects themselves are understood; it is the interaction with the integrated paraphernalia assemblage, a complex of objects that are curated indefinitely and perpetually remade, that anchors collective identities and structures corporate prestige. The theoretical framework of bundling counters the prevailing tendency to view the assembly and disassembly of ritual objects as simply symbolic or a reflection of social relations. This is particularly useful in areas like the Pueblo

Southwest where ethnographic comparisons have allowed archaeologists to interpret ritual paraphernalia and exotic trade goods as an index of prestige without addressing how prestige is created and negotiated.

The biographical perspective of theories like bundling and enchainment is particularly well suited to analyses of ritual paraphernalia. This does not make bundling the final answer to theorizing human interaction with the material. If the test of a theory is its utility in practice, tracing the social impacts of historic associations between objects is going to be unhelpful for many classes of material culture. Ritual objects tend to have value-laden and highly specific rules that dictate the contexts of their use and exchange. This makes it easier to trace their cumulative impacts on human relations. However, not every artifact has a biography that is visible archaeologically—or for that matter, one that was significant to the social contexts of its use.

There are in fact many sorts of artifacts that are not "fetishes" but that people nonetheless fetishize, in the sense that the object anchors identities and standards of behavior and is invested with a significance that impacts historical outcomes. For instance, the variable associations between ancestral territory and personal identity produced by the choice of materials in vessels for serving food is an activity that is at once prosaic and ritualized (see Coelho, Chiykowsi, and Fullen this volume). Moreover, in many cultures the sacred is not a distinctly bounded sphere but rather interdigitates with numerous aspects of everyday life, meaning that otherwise mundane objects can become invested with perceived potency by the contexts in which people use them (see Can, this volume). In such cases, concepts like bundle, enchainment, and secondary agent are fruitfully applied.

In the end, the strength of theories like bundling and enchainment is in the fact that they are nuanced enough that they can be usefully applied to certain types of archaeological analyses but not to others. As social scientists, we should be suspicious of theories that seek to provide a universal framework for analysis and instead embrace the perspectives that allow us to solve particular problems well.

Acknowledgments

Many thanks to my cocontributors to this volume for the many hours of insightful discussion and helpful commentary. Thanks also to Ruth Van Dyke for mentorship and many rounds of editorial advice and Kelley Hays-Gilpin and the Museum of Northern Arizona for making this research possible.

References Cited

Akins, Nancy J. 2001. Chaco Canyon Mortuary Practices and Archaeological Correlates of Complexity. In *Ancient Burial Practices in the American Southwest: Archaeology, Physical Anthropology, and Native American Perspectives*, edited by Douglas R. Mitchell and Judy L. Brunson-Hadley, pp. 167–191. University of New Mexico Press, Albuquerque.

———. 2003. The Burials of Pueblo Bonito. In *Pueblo Bonito: Center of the Chacoan World*, edited by Jill Neitzel, pp. 94–106. Smithsonian Books, Washington, DC.

Bernardini, Wesley. 2008. Identity as History: Hopi Clans and the Curation of Oral Tradition. *Journal of Anthropological Research* 64:483–509.

Brown, Gary M., Thomas C. Windes, and Peter J. McKenna. 2008. Animas Anamensis: Aztec Ruins or Anasazi Capital? In *Chaco's Northern Prodigies: Salmon, Aztec, and the Ascendency of the Middle San Juan Region after AD 1100*, edited by Paul F. Reed, pp. 231–250. University of Utah Press, Salt Lake City.

Cushing, Frank H. 1988 [1884]. *The Mythic World of the Zuni*, edited and illustrated by Barton Wright. University of New Mexico Press, Albuquerque.

Downum, Christian E. 2004. Dating Wupatki Pueblo: Tree Ring Evidence. Excerpted and adapted from *An Architectural Study of Wupatki Pueblo (NA 405)*, by Christian E. Downum, Ellen Brennan, and James P. Holmlund. Northern Arizona University Archaeological Report 1175, Flagstaff, Arizona. Electronic document http://jan.ucc.nau.edu/~d-antlab/Wupatki/tree ring.htm, accessed January 22, 2012.

Elson, Mark D., Michael H. Ort, Jerome Hesse, and Wendell A. Duffield. 2002. Lava, Corn, and Ritual in the Northern Southwest. *American Antiquity* 67(1):119–135.

Ferguson, T. J., Kurt E. Dongoske, and Leigh J. Kuwanwisiwma. 2001. Hopi Perspectives on Southwestern Mortuary Studies. In *Ancient Burial Practices in the American Southwest: Archaeology, Physical Anthropology, and Native American Perspectives*, edited by Douglas R. Mitchell and Judy L. Brunson-Hadley, pp. 9–26. University of New Mexico Press, Albuquerque.

Ferguson, T. J., and Micah Loma'omvaya. 2011. Nuvatukya'ovi, Palatsmo Niqw Wupatki: Hopi History, Culture, and Landscape. In *Sunset Crater Archaeology: The History of a Volcanic Landscape. Prehistoric Settlement in the Shadow of the Volcano*, edited by Mark Elson, pp. 143–186. Center for Desert Archaeology Anthropological Papers No. 37, Center for Desert Archaeology.

Gell, Alfred. 1998. *Art and Agency: An Anthropological Theory*. Oxford University Press, Oxford.

Gruner, Erina. 2012. *Post-Chacoan Ceremonial Societies on the Chaco Periphery*. MA thesis, Department of Anthropology, Binghamton University-SUNY.

Heitman, Carolyn, and Stephen Plog. 2005. Kinship and the Dynamics of the House: Rediscovering Dualism in the Pueblo Past. In *A Catalyst for Ideas: Anthropological Archaeology and the Legacy of Douglas Schwartz*, edited by Vernon L. Scarborough, pp. 69–100. School of American Research Press, Santa Fe.

Hohmann, John W. 1982. *Sinagua Social Differentiation: Inferences Based on Prehistoric Mortuary Practices*, The Arizona Archaeologist No. 17. Arizona Archaeological Society, Tucson.

Kuwanwisiwma, Leigh J. 2004. Yupkoyvi: The Hopi Story of Chaco Canyon. In *In Search of Chaco: New Approaches to an Archaeological Enigma*, edited by David Grant Noble, pp. 41–47. School of American Research Press, Santa Fe.

Latour, Bruno. 2005. *Reassembling the Social: An Introduction to Actor-Network Theory*. Oxford University Press, Oxford.

Lekson, Stephen H. 2008. *A History of the Ancient Southwest*. School for Advanced Research Press, Santa Fe.

Levy, Jerrold E. 1992. *Orayvi Revisited: Social Stratification in an "Egalitarian" Society*. School of American Research Press, Santa Fe.

McGregor, John C. n.d. Burial of an Early American Magician. Unpublished notes and manuscript on file at the Museum of Northern Arizona, Flagstaff.

———. 1941. *Winona and Ridge Ruin, Part I: Architecture and Material Culture*. Museum of Northern Arizona Bulletin 18. Northern Arizona Society of Science and Art, Flagstaff.

———. 1943. Burial of an Early American Magician. *Proceedings of the American Philosophical Society* 82(2):270–298.

Mills, Barbara J. 2004. The Establishment and Defeat of Hierarchy: Inalienable Possessions and the History of Collective Prestige Structures in the Pueblo Southwest. *American Anthropologist* 106(2):238–251.

Mills, Barbara J., and T. J. Ferguson. 2008. Animate Objects: Shell Trumpets and Ritual Networks in the Greater Southwest. *Journal of Archaeological Method and Theory* 15:338–361.

Neitzel, Jill E. 2003a. Three Questions About Pueblo Bonito. In *Pueblo Bonito: Center of the Chacoan World*, edited by Jill Neitzel, pp. 1–9. Smithsonian Books, Washington, DC.

———. 2003b. Artifact Distributions at Pueblo Bonito. In *Pueblo Bonito: Center of the Chacoan World*, edited by Jill Neitzel, pp. 107–126. Smithsonian Books, Washington, DC.

O'Hara, Michael. 2008. The Magician of Ridge Ruin: An Interpretation of the Social, Political, and Ritual Roles Represented. Paper presented at the 73rd Meeting of the Society for American Archaeology, Vancouver.

Ortiz, Alfonso. 1994. The Dynamics of Pueblo Cultural Survival. In *North American Indian Anthropology: Essays on Society and Culture*, edited by Raymond J. DeMalle, pp. 307–327. University of Oklahoma Press, Norman.

Parsons, Elsie Clews. 1996 [1939]. *Pueblo Indian Religion*. 2 vols. University of Nebraska Books, Lincoln.

Pauketat, Timothy R. 2013a. Bundles of/in/as Time. In *Big Histories, Human Lives: Tackling Problems of Scale in Archaeology*, edited by J. Robb and T. Pauketat, pp. 35–56. School for Advanced Research Press, Santa Fe.

———. 2013b. *An Archaeology of the Cosmos: Rethinking Agency and Religion in Ancient America*. Routledge, New York.

Pepper, George H. 1906. The Exploration of a Burial Room in Pueblo Bonito, New Mexico. In *Anthropological Essays: Putnam Anniversary Volume*, by his friends and associates, pp. 196–252. G. E. Steckert, New York.

———. 1996 [1920] *Pueblo Bonito*. University of New Mexico Press, Albuquerque.

Reed, Paul F., (editor). 2008. *Chaco's Northern Prodigies: Salmon, Aztec, and the Ascendency of the Middle San Juan Region After AD 1100*. The University of Utah Press, Salt Lake City.

Reid, J. Jefferson, and Stephanie Whittlesey. 1997. *The Archaeology of Ancient Arizona*. The University of Arizona Press, Tucson.

Schillaci, Michael A. 2003. The Development of Population Diversity at Chaco Canyon. *Kiva* 68(3):221–246.

Stanislawski, Michael. 1963. *Wupatki Pueblo: A Study in Cultural Fusion and Change in Sinagua and Hopi Prehistory.* Unpublished PhD dissertation, Department of Anthropology, University of Arizona. University microfilms, Tucson.

Stein, John, Dabney Ford, and Richard Freidman. 2003. Reconstructing Pueblo Bonito. In *Pueblo Bonito: Center of the Chacoan World*, edited by Jill Neitzel, pp. 33–60. Smithsonian Books, Washington, DC.

Stevenson, Matilda Coxe. 1904. *The Zuni Indians: Their Mythology, Esoteric Fraternities, and Ceremonies.* Twenty-third Annual Report of the Bureau of American Ethnology to the Secretary of the Institution. Government Printing Office, Washington, DC.

Suderman, Mrs. John P. ca. 1945. *Our Missions Among the Hopi Indians.* Notes on file at the Mennonite Library and Archives, Bethel College, North Newton, Kansas.

Titiev, Mischa. 1968 [1944]. *Old Oraibi.* Peabody Museum of American Archaeology and Ethnology Volume XXII, No. 1. Harvard University Press, Cambridge.

Toll, H. Wolcott. 2006. Organization of Production. In *The Archaeology of Chaco Canyon: An Eleventh-Century Pueblo Regional Center*, edited by Stephen H. Lekson, pp. 117–152. School of American Research Press, Santa Fe.

Van Dyke, Ruth M. 2004. Memory, Meaning, and Masonry: The Late Bonito Chacoan Landscape. *American Antiquity* 69(3):413–431.

———. 2009. Chaco Reloaded: Discursive Social Memory on the post-Chacoan Landscape. *Journal of Social Archaeology* 9(2):220–248.

Vivian, R. Gwinn, Dulce N. Dodgen, and Gayle H. Hartmann. 1978. *Wooden Ritual Artifacts from Chaco Canyon, New Mexico: The Chetro Ketl Collection.* Anthropological Papers of the University of Arizona No. 32. The University of Arizona Press, Tucson.

Vokes, Arthur W., and David A. Gregory. 2007. Exchange Networks for Exotic Goods and the Zuni's Place in Them. In *Zuni Origins: Toward a New Synthesis of Southwestern Archaeology*, edited by David A. Gregory and William H. Doelle, pp. 318–360. University of Arizona Press, Tucson.

Webster, Laurie D. 2011. Perishable Ritual Artifacts at the West Ruin of Aztec, New Mexico: Evidence for a Chacoan Migration. *Kiva* 77(2):139–171.

White, Leslie A. 1932. *The Acoma Indians.* Extract from the Forty-Seventh Annual Report of the Bureau of American Ethnology. United States Government Printing Office, Washington.

———. 1962. *The Pueblo of Sia, New Mexico.* Bureau of American Ethnology Bulletin 184. United States Government Printing Office, Washington, DC.

Whiteley, Peter. 1998. *Rethinking Hopi Ethnography.* Smithsonian Institution Press, Washington.

Windes, Thomas C. 2003. This Old House: Construction and Abandonment at Pueblo Bonito. In *Pueblo Bonito: Center of the Chacoan World*, edited by Jill E. Neitzel, pp. 14–32. Smithsonian Books, Washington, DC.

Zedeño, Maria Nieves. 2008. Bundled Worlds: The Roles and Interactions of Complex Objects from the North American Plains. *Journal of Archaeological Method and Theory* 15:362–378.

CHAPTER FOUR

Animacy of the Everyday

Materiality, Bundling, and the Production of Quotidian Ceramics

Tanya Chiykowski

The metamorphosis of clay into a vessel is a fundamentally transformative process; at each stage potters capture and manipulate the spirit of the pot through their interactions with it. The vessel that results is a bundle of these actions dictated by the fictile and malleable qualities of the clay and the actions of the potter. Kopytoff (1986:68) introduces object biographies, which examine objects as culturally constructed entities endowed with culturally specific meanings; consequently, archaeologists have unique insights into the physical processes of object construction. Viewing a finished object as a bundle links materials and experience, as "object biographies comprise life histories, genealogies of practices, and histories of place that define larger fields of social relationships" (Pauketat 2013:29).

Each stage of ceramic production bundles traits, which creates an animate finished pot. Arguments for the animacy of effigy vessels (VanPool and Newsome 2012) and polychrome ceramics (Charley and McChesney 2007; Walker and Burt 2009) rest on the process of manufacture, not just in the effect of the object. In the American Southwest, ethnographers (Bunzel 1972 [1929]; Stevenson 1904) recorded potting traditions; these documents include the informant's emphasis on not just the physical steps of production but the spiritual implications as well (Martínez 1988). If it is the original clay and process of transformation that bring these objects to life, then the same principles apply to a wide range of material culture, including quotidian plainware ceramics, which make up the vast majority of archaeological assemblages.

In this chapter, I explore the implications of animate plainwares at the site of Cerro de Trincheras, Sonora, Mexico. Cerro de Trincheras was a large, terraced hilltop village occupied by Trincheras tradition populations from AD 1300 to 1450 (McGuire and Villalpando 2011). The Trincheras tradition is the southernmost extent of the greater Southwestern cultures. In prehistory, information, goods, and spiritual values flowed between Trincheras populations and other Southwest cultures, such as the Hohokam, Mogollon, and Ancient Pueblo groups. Ethnographers, such as Fontana (1962) and Lumholtz (1912), documented the practices and beliefs of potters in the Altar Valley, located fifty kilometers from the abandoned site of Cerro de Trincheras. A common idea was that nonhuman entities could be animate.

Animacy resulting from material qualities is always mediated via the human experience and makes material agency secondary to the agency of humans (Gell 1998:74). The materiality of clay and its sensuous, affordant qualities (Pauketat 2013) suggests the widespread conviction that the process of ceramic production is also a spiritual transformation (Boivin 2008; Vitelli 1999; Wengrow 1998). Pan-Amerindian beliefs reject the detachment of natural/supernatural and the separation of the two into segregated locations, such as shrines; instead, natural and supernatural are in constant coexistence in all places (Viveiros de Castro 1992, 1998; Weismantel 2013). At the archaeological site of Cerro de Trincheras, plainware ceramics provide evidence for how spiritual practices were enacted in daily activity through the production of visually "mundane" goods (Stinson 2010). Archaeologists then interact with sherds in a markedly different way; the transformation for us rests on the inherent antiquity of the items. The transformation of clay to pot to archaeological context imbues the artifact with significance that causes conflict over cultural patrimony between modern indigenous nations and nonnative governments. Once re-animated and reintegrated into social contexts, sherds become part of the power struggle over ownership of the past.

Why Pots?

In order to discuss the deep past without ethnographic references, archaeologists have relied on the material qualities of the object. Once transformed from clay, ceramics are durable but fragile. Unlike other materials, such as wood or stone, the process of creating a vessel fundamentally and irreversibly changes the properties of the clay and makes the history of its production critical to understanding the product (Wengrow 1998).

Kopytoff's (1986) concept of object biographies demonstrates how objects become imbued with meaning through their life histories and interactions with people. Archaeologists are familiar with the framework of object biographies from *chaîne opératoire* and use-life studies (Lechtman 1977; Lemmonier 1986; Schiffer 1999), which examine the production, use, and reuse of an object, not just its final deposition. Kopytoff (1986) and Appadurai (1986) move anthropologists away from studying production and exchange and toward the roles of the objects themselves. The value of the object is not limited to its function or equivalent worth, but it encompasses the social relations it represents. Artifacts are not abstract goods where one widget can stand in for another; rather the object and its qualities are part of ongoing relationships that embroil people and things (Thomas 1996:159). Although many examples of life histories of objects focus on the circulation of finished products, studying the process of production allows us to follow the way each step forms and bundles traits, creating an entangled finished vessel. When Fontana (1962) described Altar Valley potters' traditional production methods, he organized his observations based on their understanding of the major production steps. These steps intertwine material and process in a seen pattern across the Southwest and are often described as the birth of pots (Charley and McChesney 2007).

This entanglement of ceramics and people goes beyond finished vessels (Hodder 2012). Chapman (2000) emphasizes the key material difference of pottery over previous materials, such as woven baskets, that is the breakability of the vessel. Once shattered, the vessel can never be completely repaired. The purposeful breaking of vessels that symbolize relationships represents a broken or severed connection that cannot be made whole again. Similarly, an association between object and place may extend beyond the use life of the original vessel. Repairing a cracked pot, curating the sherds, or depositing them in a cache all reflect the added social value placed on the ceramics as symbols of relationships, even after the object has ceased to perform its primary function.

Thomas (1996:162) discussed how through creating and trading objects the item exchanged becomes a stand-in for relationships. It is for this reason that ceramic sourcing studies can be of value. Rather than tracking the long distance trade of objects, we focus on the series of relationships that form them. This becomes a way of humanizing the biplot graphs of INAA (Instrumental Neutron Activation Analysis) and the point counts of petrography. If these objects are entangled with the relationships that literally formed them, then archaeologists can expect to see how people used these objects in new ways. Ceramic objects then are bundles (the coming together of people and materials) made whole by the embodied experiences and material practices of a society (Pauketat 2013). A sherd is a representation of a whole vessel, which tangles together raw resources (clay), technology (forming and firing of vessels), and social relationships ([learning and exchange] Hodder 2012).

The following case studies demonstrate fruitful avenues for exploring how clay and ceramics are active forces in society, and they extend Kopytoff's object biographies to all stages of manufacture, not just the circulation of finished objects. I apply this methodology to plainware ceramics from the site of Cerro de Trincheras (Figure 4.1). Objects have powerful roles in domestic life, as well as in formal sacred environments (Stinson 2010). Demonstrating animacy in quotidian ceramics can only underscore

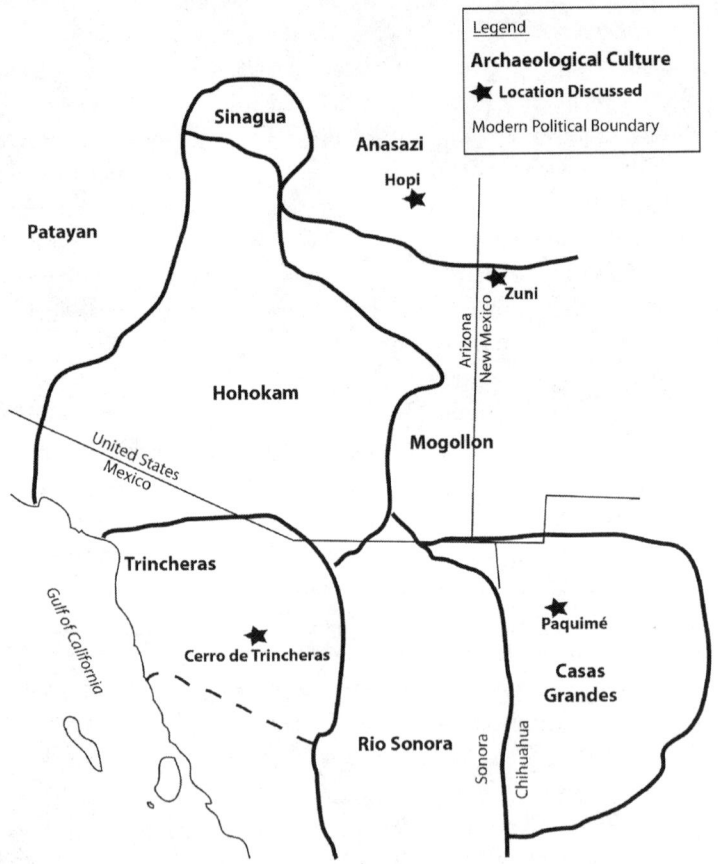

Figure 4.1. Map of the greater Southwest and locations discussed in the text. Drafted by Tanya Chiykowski based on McGuire and Villalpando (1993).

the entanglements exhibited by more exotic counterparts, such as effigies (VanPool and Newsome 2012) and polychromes (Walker and Burt 2009).

The Transformative Power of Clay

Archaeologists studying materiality in Neolithic contexts around the world have begun to discuss the properties of clay. Frequently, it is the materiality of the objects while being formed that creates the possibility of the finished vessel becoming a social agent. A common thread is the awe attached to the transformation of raw material to the finished vessel—an observation that is present in ethnographic conversations with Hopi and Zuni informants (Bunzel 1972 [1929]; Charley and McChesney 2007; Martínez 1988; Parsons 1996 [1939]; Stevenson 1904).

Recent approaches to ceramics have incorporated more explicit materiality and object agency. Many of these cases focus on the Neolithic and the transition to sedentism. Ceramics are one of many objects in the "Age of Clay" (Wengrow 1998), as people built walls and hearths surrounding themselves in clay. The properties of clay, including its underground origins, the process of wetting, mixing, building, drying, firing, and disintegration all had important psychological impacts on the populations using the materials. This dependence on clay developed in sync with increased awareness of soils and investment in the earth.

Although Vitelli (1999) did not use the terms materiality or object agency, her description of early Greek ceramics focuses on the transformative process of clay and how early leaders may have used the power of this transformation. In her analysis, the functional benefits of Neolithic clay pots from southern Greece fall short of explaining the adoption of pottery. The poorly fired pots would not have held liquid for long and were generally too small to have been storage devices for dry goods. Additionally, there is a long period of time when they persist in small quantities without growing in popularity. Potters made on average twelve to thirteen pots a year. Yet for four hundred years, physiochemical analysis of the pots demonstrates that there were four temper recipes for making the pots that remained distinctive. This indicates that the method of construction remained restricted to a few producers over this long period of time. Owners of the pots mended many of them when they broke, despite a lack of use wear on the vessels.

Functional explanations of ceramic adoption fail in this case; Vitelli (1999) focuses on the novel aspects of ceramics. People (in her argument predominantly women) were familiar with the properties of fired clay from building hearths. Shamans also tended to use earth substances for healing purposes (Vitelli 1999). The process of a shaman transforming wet clay into hard containers would have been a performative act. The fire and smoke would have emphasized the metamorphosis of material through the consumption of wood fuel into light. Once formed, the vessels would have been tokens of an experience, mementos of the time and place of the ritual that formed them.

Vitelli supports this argument with the analysis of calcium carbonate in the early pots. This has undesirable effects on the finished product by making it crack and crumble relatively easily. Potters in pre-seventh-millennium, southern Greece deliberately added crushed calcite and limestone as temper. If the goal was a durable pot, this is a counterproductive practice. If, however, it was the process of forming and firing the vessel that was important, the disintegration of the final product may have been part of the magic.

What Vitelli (1999) presents is a coherent, cohesive argument for process over product. Although working in the deep past limits the archaeologist's ability to make definitive claims, Vitelli's (1999) work in the Neolithic provides a new way of thinking about early ceramics that moves beyond functional adaptations to focus on the transformative properties of making a pot.

Boivin (2008) also has written extensively on this topic from the vantage point of phenomenology and material culture studies. Her inspiration came from ethnoarchaeological interviews with Rajasthani villagers in India. The villagers see certain red soils as the physical manifestation of the goddess Laksmi. Boivin realized that non-Western populations conceptualized the earth in very different ways than early agricultural populations. The ability to manipulate and transform clay resonates with how Neolithic populations manipulated soils and the landscape as unprecedented modification of the natural world.

Melted clay from walls made up a significant portion of the tells on which Neolithic populations continued to live. These walls would have needed continual upkeep in an attempt to renew and revive the buildings (Hodder 2012; Wengrow 1998). It is this revitalization and reanimation of architecture that created the complicated micromorphology of walls in Neolithic sites (Boivin 2008). At times, revitalizing the walls was not enough; learning from the properties of fired ceramics, Neolithic populations in southern Europe burned their houses when abandoning them. The result was a durable mark on the landscape that maintained a claim to place (Stevanovic 1997). As agricultural populations invested more in their soils for agriculture, they used the earth as a means of marking distinction.

Alongside this modification of soils came an increase in the use of clays for symbolic communication. Clay is fictile; unlike other materials, such as stone or wood, it has no grain and can be made into any shape (Wengrow 1998). This malleable nature allowed for new symbolic forms seen in the Neolithic. Clay figurines, seals, and decorated ceramics all became a means to communicate identity. Conversely, Hodder (2011, 2012) suggests that at Çatalhöyük the opposite happened; due to their functional characteristics, pots were entangled in the world of domestic production and specifically not chosen for their symbolic potential. As such, the Levant IV ceramics remain largely undecorated and are not included in burials or ritual contexts.

The Neolithic case studies provide insight into how an emic perspective of a subset of material culture changed with the advent of agriculture and sedentism. However, the conclusions of Hodder (2011) and Wengrow (1998) demonstrate the difficulty of applying materiality in past contexts, especially when there is a lack of ethnohistoric context. Both arguments appeal to the qualities of the material and seem plausible; yet there is no way to evaluate or verify either assumption. Ethnographic data provides a wealth of information that demonstrates the validity of materiality approaches by using an emic perspective of the culture understudy. These viewpoints are important, especially in the greater Southwest where object animism breaks down the division between the natural and supernatural worlds (Mills and Ferguson 2008).

Animacy from Clay: Southwestern Decorated Ceramics

Earlier work on materiality took a rigorous look at the interplay between pot and potter to explain how person and object influence each other; the prototype of the pot influences the artistic expectations of the potter (Gell 1998:74; Fullen this volume). Using an example of a Hopi pot, Schiffer (1999) discusses this in a roundabout way but admits that a pot can be a "sender" in his schema of behavioral archaeology. Here the artist becomes the receiver of the half-painted design on the pot, and his/her reaction and design is influenced by this message. The artist is not unaffected by the influence of the pot, and she or he will change the scheme to meet the expectations of the design when entering into a dynamic relationship between person and object (VanPool and Newsome 2012).

The European Neolithic examples rely on physical properties of clay, but in southwestern North America archaeologists can build arguments of animacy from extensive ethnographic and ethnohistoric records from groups such as Hopi and Zuni. Early ethnographers, including Stevenson (1904) and Bunzel (1972 [1929]), collected information on how Pueblo women made their pots. They recorded not just the production sequence but also the values and beliefs associated with the transformative process often described using birth metaphors (Charley and McChesney 2007). These beliefs are widespread across the greater Southwest and are detailed by potters, such as María Martínez of San Ildefonso (Martínez 1988). Fontana's description of Sonoran potters contains many parallels, and he describes pottery creation as a female activity completed in the home. Within this domestic context the physical skills, technical knowledge, and ideological importance were passed on, learned, and perpetuated by immersive education. I use these ethnographic analogies not to argue that Trincheras populations were identical to modern Hopi potters but to discuss the operationalization of animacy across the greater Southwest.

It is important to note that for Southwestern groups objects made of clay are animate, but they are not necessarily acting social agents. Mills and Ferguson (2008) demonstrate how animacy bridges the Western divide between the natural and supernatural. Archaeology becomes a powerful means of accomplishing this, linking the role that physical objects play with spiritual practices. Researchers often highlight extraordinary objects, such as shell trumpets and effigy vessels. However, appeals to animacy all start from the same place—the properties of the clay. María Martínez (1988) discusses the offerings of cornmeal and spiritual requirements for removing clay. This material is the body of Grandmother Clay or Old Woman Clay, who lets potters use it (Parsons 1996 [1939]:195). By contrast, other materials, such as temper and water, lack similar strictures and rituals around their use (Martínez 1988).

Ethnographic conversations with modern informants provide evidence of the enduring dispositions linking the power of raw clay to the finished product of a pot. Hays-Gilpin (2011) recognized that most collections of Hopi pottery represent the efforts of Hopi potters producing for the art market or for ethnographic museum curation. Despite artists' assertions that these objects are meant to be active parts of social life, ceramics as art objects have changed the ways in which people view them. Out of its cultural context, the perfection of the pot causes a different response by the owner. The pots are now too fragile to use. They must be protected from scrapes and bumps, and the collector is required to remove the pot from its role of tool where it is meant to be an active part of society. To the collector, any damage or breakage abruptly terminates the value of the vessel. Archaeologists generally assume sherds are the inert refuse of long dead people. However, for indigenous peoples, sherds may be active markers of ancestors on the landscape (Colwell-Chanthaphonh and Ferguson 2006).

Hays-Gilpin (2011) shows how museum curators and Hopi artisans deliberately cultivated an association between ancient pots and modern craft. In the early twentieth century, Museum of Northern Arizona personnel copied designs from Hopi artifacts in collections and suggested that modern Hopi artisans use the iconography to differentiate their art from other native groups. The materiality of museum collections created a definition of Hopi art. As similar processes occurred with Hopi production, museum personnel pressured artists to use what curators' deemed traditional techniques, such as coal firing. Perfectly fired vessels go against the belief that open, dung fired ceramics have "life" that is visible through the mottled, uneven color of the firing surface (Charley and McChesney 2007).

VanPool and Newsome (2012) build on approaches to object agency in North American archaeology by using culturally specific information to build a case of object animacy at the site of Paquimé. Effigy vessels represented specific people, broken hand-drum sherds sanctified space, and pots with jewelry were adorned like humans. Using Pueblo ethnography and the depositional context of ceramic objects, VanPool and Newsome (2012) demonstrate that ritual vessels had a spirit that made them animate. The vessels' spiritual characters derived from the physically transformative power of clay. This transformation of clay into object could also further sanctify spaces even when broken, as in the case of hand drums. The special treatment of ceramic objects by Paquiméans demonstrates the respect and reverence the objects had because of their aliveness and ability to affect society.

A Biography of Plainware Pots

As materiality and object agency approaches gain currency in archaeology, discussions are moving from ritual to quotidian contexts. Like ritual objects, utilitarian items may also have bundled traits; in fact, the separation of space into ritual and domestic, and objects into mundane and spiritual, is an artificial dichotomy (Fowles 2013; Stinson 2010). In the above examples, the animacy of ritual ceramics stems from the spiritual qualities of the raw clay. Potters form ceramics through a series of stages, each of which imparts nontangible qualities to the final pot. Although the effects of the pot are most noticeable in elaborate vessels, the same actions create plainware vessels with similar qualities. It is the process of making the pot, not the vessel, in and of itself, that animates the object. The object biographies of animate, quotidian vessels can provide insight into everyday beliefs and practices.

My case study involves ceramics from the late prehistoric Trincheras tradition (see Figure 4.1 above) of southern Arizona and northern Sonora (McGuire 2008; McGuire and Villalpando 1993, 2011; Villalpando 2011). During approximately AD 1300 to 1450 people in the region aggregated in terraced hilltop villages, the largest of which was the site of Cerro de Trincheras. In earlier periods potters had decorated ceramics with red designs and purple specular paint. However, in the late prehistoric virtually all of the locally produced pottery was undecorated plainware, including the vessels used in cremation burials (Figure 4.2). At Cerro de Trincheras, plainwares comprised over a million sherds and 99 percent of the ceramics. By contrast, there were very few decorated ceramics, and these were mostly imported from regions such as Paquimé (Gallaga 2011). The emphasis on domestic production reorients discussion of animacy away from restrictive ritual contexts. The false dichotomy of domestic and sacred contexts breaks down when discussing the animating forces of clay (Stinson 2010). All evidence suggests female potters at Cerro de Trincheras produced their quotidian ceramics as part of domestic life; to ignore the majority of sherds is to overlook a continuous and omnipresent form of material culture and the associated spiritual value of these artifacts.

The Trincheras tradition of northern Sonora lacks modern analogs with groups, such as the Hopi and Zuni, who can provide a direct historical approach. Contemporary Tohono O'odham populations live nearest to Cerro de Trincheras and are regularly consulted on archaeological plans and findings (McGuire 2008:171); however, they have few specific traditions that specifically discuss Cerro de Trincheras ceramics. Nonetheless, the Trincheras tradition is part of the greater Southwest where groups exchanged ceramics and ideas across cultural boundaries for millennia. I use O'odham (formerly referred to as Pima and Papago groups) and Pueblo ethnography, as they are the most relevant comparisons available. Many of the points I draw from ethnography come from several ethnohistoric and contemporary Pueblos, and they represent widely held beliefs. I break down the production process into standard stages: collecting materials, forming the vessel, decoration, firing, use, discard, and collection. Potters around the world recognize these stages, and Southwest groups discuss

Figure 4.2. Plainware cremation vessel from Cerro de Trincheras. Photograph used by permission of INAH Sonora, photo taken by Randall McGuire and María Elisa Villalpando.

them in the ethnographies I cite. This approach illustrates how the qualities of each stage build to the animacy of the final object. The same beliefs and forces that animated polychromes for the Mimbres (Walker and Burt 2009) and Paquiméans (VanPool and Newsome 2012) were played out in the manufacture of quotidian ceramics at Cerro de Trincheras.

Collecting Materials—Clay Sources

From the very beginning of the ceramic production process, the earth is seen as sacred. Like Boivin's (2005) findings that red clay is the god Laksmi, North American groups find parts of the earth to be alive. As previously noted, in many Southwest traditions, clay is the flesh of Grandmother Clay or Old Woman Clay, who lets potters use it (Parsons 1996 [1939]:195). Because clay is the body of Grandmother Clay, it is associated with females and there are prohibitions against men working with clay. Traditionally these gendered requirements by the clay mean potters are female, but the effect extends beyond just vessels; women complete the plastering of houses and kivas ([although men paint them] Parsons 1996 [1939]:38).

Through conversations with her informant, We'wha, Stevenson (1904:374) found that the gendered nature of resources applied to the collection of clay. We'wha was a transgendered individual who took on female roles and dress; she had access to male and female ritual information and the power associated with it. When collecting clay with Stevenson, We'wha asked the cameraperson to stay away from the clay bed because he was male, and it was therefore dangerous. As We'wha started collecting the clay, she told those around her to not ask questions or talk because praying to the clay took her complete attention. These prayers were necessary to properly extract the clay, since it was in a sense alive.

The clay resources can be entangled in human relations in other ways as well. Not all clay is equal. The Zuni obtain clay from one specific source, Corn Mountain (Stevenson 1904:374). Cross culturally potters do not stray far from their home base for clays (Arnold 1991). However, Zuni potters demonstrate that technical concerns are only one part of the resource selection decision. Potters continued to use one specific source despite there being a location closer to the community; they said this source replenished itself and would never run out. Although the whole world may be alive, specific places have characteristics that make some materials more powerful than others. Fontana (1962:23) recorded stories of O'odham potters traveling tens of kilometers specifically to make pots in the Altar Valley (fifty kilometers north of Cerro de Trincheras). Their journey underscored the emphasis on place and landscape on the finished vessel. For this reason, people must take spiritual care when visiting these locations and removing resources. The finished product will keep the spiritual power associated with the place where it was collected.

Forming the Vessel

Modern hobby and commercial potters follow specific detailed recipes for mixing the correct ratio of temper to clay. This kind of formulaic understanding was not the process of preparing and forming the vessels in the past because resources were so variable. Ethnographic interviews with potters highlight the importance of feel when it comes to the correct ratio of clay, temper, and water (Bunzel 1972 [1929]:6). It is an evaluation process rooted in the tactile experience of the clay. In Sonora the reliance on texture would have been particularly salient, as the redeposited secondary clays would be heterogeneous even within the same source. As populations and potters moved around, the ability to feel for the correct texture would have been important as the clay sources changed. Knowing through touch how the clay should feel would allow potters in a new area to continue their craft. Suitability of the clay is not limited to its tactile properties. Fontana (1962:118) notes the importance of taste for O'odham potters when they make pots. Bitter clays contain salts, which decreases durability and creates scumming on the surface of the vessel (Fontana 1962:68). Preparation of the vessel form was a whole body experience, drawing in all of the potter's senses.

In Sonora around AD 1300 there was also a change in the technology of production when Hohokam populations introduced paddle-and-anvil manufacture. At the site of Cerro de Trincheras potters used the traditional coil-and-scrape and paddle-and-anvil methods concurrently; in rare cases potters used both methods to form one vessel (Gallaga 2011). How manufacturing differences would affect the materiality of production is debatable, as there is minimal research in this area. Archaeologists can shift their analysis to a smaller scale to study the role of the physical, kinetic experience in everyday activities. Older O'odham potters in southern Arizona describe the experience of waking up to the soft hammering of paddle-and-anvil ceramics (Fontana 1962:13; Lumholtz 1912:16). This would have noticeably changed the soundscape of Cerro de Trincheras, which naturally forms an amphitheater-like bowl where sound echoes (O'Donovan 1997). Additionally, potters making paddle-and-anvil vessels would have had different space requirements and movement patterns that would have conditioned how they went about their potting routines. Experimental archaeology, watching traditional potters, and re-creating ethnographic methods can all contribute to materiality-based projects.

The adoption of paddle-and-anvil technology at Cerro de Trincheras represents a greater disconnect than a simple altering of methods. Hays-Gilpin (2008) points out the conceptual parallels between the coil-and-scrape production sequence and earlier basketry (Parsons 1996 [1939]:38). There are mechanical and visual equivalencies between the two that would be severed with paddle-and-anvil methods. Would the finished vessel have meant something different because it was formed in a new way? Regardless, the formation is where the pot is "born." In this way, forming a pot is similar to nurturing a child; the pot can carry forward the values of the potter, so care must be taken not to transfer the wrong emotions to the pot (Charley and McChesney 2007).

Decoration

Frequently, discussions of decoration on ceramics become about meaning and what the symbols represent. At Acoma and Hopi, where potters believe that each pot is an individual, prohibitions against perceived copying of designs protect the unique identity of each pot (Bunzel 1972 [1929]:52). Often the medium of expression is inconsequential to the effect. Some authors in the Southwest are starting to move beyond that by focusing on how the form of the vessel influences interaction with design (e.g., Crown 1995; Van Keuren 2006; VanPool and VanPool 2007). Factors like the undulation of snakes or the radial patterns of neck decorations when viewed from above are all experiences that depend on the form and manifestation of the pot.

Ethnographically there are important ways in which the power of pots manifested itself in the decoration. Hays-Gilpin (2008) studied the variation in closing the lifeline (the painted band around the top of an open vessel); there were differences in whether a gap was present and the extent of this gap. The material correlation between earlier basketry and the decoration on pottery may perpetuate the material constraints of earlier methods. Ethnohistorically, pregnant women, and sometimes women of childbearing years, did not close the band on the pot because it could cause difficult labor (Parsons 1996 [1939]:91). It is important to note that the agency of objects was therefore not perceived to be uniform; depending on gender, age, ritual, or status objects would have different effects.

Firing

In firing, the pot is permanently transformed from a malleable material to a solid and breakable object. This is a volatile step during which much can go wrong to affect the spirit of the pot. For example, a pregnant woman being or talking too close to the fire causes black spots, or fire clouds. To mitigate these risks, potters may make offerings to the fire or have the kiln ground blessed before beginning the firing process (Parsons 1996 [1939]:293). Sometimes this offering was corn bread so that the vessel's spirit could be fed, again showing how the formed pot is consuming and alive.

This point of transformation is not always dangerous; it has a positive power as well. Not all variation created in the manufacturing process is a bad thing. The variegated surface created by traditional pit firing is something that decreases the perfection for art collectors, but it is what gives the pot life. This "blush" is the spirit of the pot in action; kiln-fired pots, despite their perfection, are "dead" because they lack this indication of a life force (Charley and McChesney 2007). The importance of dappled surfaces has repercussions for understanding that all pottery has a spirit, not only anthropomorphic vessels or elaborate polychromes. At Cerro de Trincheras firing clouds are common on ceramics and they produce tremendous variation in surface color (Gallaga 2011). Additionally, there is no evidence for special-purpose kilns, so these firing rituals were performed in a domestic context, sanctifying mundane space. Within the home these values would have been perpetuated as kinswomen waited to see the results of their efforts.

The visual qualities are not the only way that a pot expresses its wholeness. Sound is an important consideration. Even if the finished pot is visually without cracks or fissures, it is not complete unless it rings when struck. This is the voice of the clay, and resonation in the shape and wholeness is what makes it complete (Charley and McChesney 2007; VanPool and Newsome 2012).

Use

Since most ceramics are found in postdepositional contexts, it is difficult to determine what role materiality played in daily activities. In Sonora, there is an almost complete lack of complete vessels. Working from sherds, archaeologists can quantify the finished pot's vessel diameter, vessel shape, volume, and fabric strength. Although useful information, these attributes provide a limited indication of how users would have interacted with the vessel.

The agency and materiality of a pot in use is tied up in what Kopytoff (1986) would call its *object biography*. The agency that is exhibited in every stage of production becomes bundled into the finished vessel. Art collections value the pristine, perfect pot, but it is the variety from its production that creates the liveliness that makes it culturally significant. People bump and scratch their products in the process of daily use and mundane activities (Hays-Gilpin 2011). The prehistoric inhabitants of Trincheras people would have interacted most frequently and would have been most familiar with utilitarian pots. In comparison, Paquimé polychrome (similar in style to the vessels discussed by VanPool and Newsome 2012) vessels were scarce; one entered the Cerro de Trincheras settlement every two to three years on average (Gallaga 2011). While the trade and exchange with Paquimé may have bundled relationships and place into a polychrome ceramic, plainwares would have enchained local behavioral traditions, which were reinforced and replicated frequently each time household potters produced their quotidian wares.

Discard

If the stages of production imbue the ceramic vessel with a spirit, what happens to that spirit when the object breaks? Cushing (1886:510) observed that the Hopi viewed the sound of a broken pot as the spirit of the pot being released. Archaeologists clearly see the idea of animate pots in the "killed" Mimbres vessels, which Brody and Swentzell (1996) argue was to liberate the breath of the pot. Marshall (1997:70) argues that the Chaco North Road is a cosmological link to the underworld emergence place of Corn Mother where people are "earthborn." At the end of this road Chacoans smashed great quantities of grayware ceramics, an activity that physically linked the release of the spirit of the pot with the return of spirits to the origin place.

Even with a broken object the power is not lost, and the values held by Mimbres and Chacoan groups remain present in modern Puebloan society. Colwell-Chanthaphonh and Ferguson (2006) detail how, for the Hopi and Zuni, sherds are part of the footprints of the ancestors. The agency placed in the sherds is still recognized and present. Sometimes the agency of the initial vessel is not left to dissipate back into the landscape. Sherd temper, while it has functional benefits, also may have important significance when it is recycled back into a new pot.

Burial of the dead in middens was a common practice in the greater Southwest, which would have contributed to the ongoing power of sherds subsequent to their deposition. Around the Cerro de Trincheras site archaeologists have recovered cremations housed in large vessels, the majority of which are plainwares (see Figure 4.2). Here, intact pots were deliberately removed from daily use to contain the remains of the dead, sanctifying the body after death. Additionally, historic O'odham groups have buried individuals in the terraces at Cerro de Trincheras (Villalpando 2011). The architectural features of the site are far from sterile but instead were permeated with sherds and other detritus of daily life. This example is the clearest evidence of the ongoing spiritual significance of quotidian ceramics.

In the Collection

As archaeologists, we remove materials from their resting places, and we reintroduce them into systems of exchange. In doing so, we start to evaluate the materiality under a new set of criteria, once again focusing on the raw material transformed into artifact. The biography of an object does not end with its collection (Hays-Gilpin 2011). A ceramic sherd is valued because it is old and because it has been recovered from an archaeological context (Holtorf 2002). Destructive analysis is often justified on plainware ceramics and body sherds because of a lowered archaeological value.

However, if we extend the argument, the same qualities that make effigy vessels animate also infuse plainwares with life.

Contemporary indigenous communities in Sonora are not overly concerned with the continued life force of sherds (McGuire 2008:171), but the effect of the ceramics continues via other qualities. A sherd's antiquity transforms the object from a piece of discarded refuse to an archaeological artifact drawing the archaeologist into a new relationship with government legislation as it becomes part of Mexican patrimony. While archaeologists and legislators no longer see the artifacts as animate in the same way that the potters did, it is the transformation from raw clay to human object that creates awe (Holtorf 2002). The awe of antiquities has led to the exportation of artifacts in colonial and imperialist archaeologies. Because of the power of the artifact, archaeology then becomes a politicized practice, especially in the borderland region of Sonora (McGuire 2008). My research occurs in a context where American academic research comes up against Mexican national patrimony and O'odham groups on both sides of the international border. As an archaeologist my actions and requests for destructive analysis continue to replicate this belief in the inherent value of sherds because of their age, and the object biography of the sherd takes on a new life stage as an object of political power.

Summary and Conclusions

Across the Southwest indigenous peoples see pots as animate because of the processes that form them. Archaeologists can engage these beliefs with an object biography approach that begins with the collection of raw materials and extends through the life of the pot. Many authors have focused on the exchange of ceramics in finished form. However, a materiality perspective takes into account the qualities of the raw materials and how they change throughout the process.

In the ancient Southwest, it is the process of production—the object biography—that gives ceramic artifacts their animacy. It is a process seen through ethnographies of the Hopi (Parson 1996 [1939]), Zuni (Stevenson 1904), O'odham (Fontana 1962), and contemporary indigenous potters (Martínez 1988). My argument highlights the properties of clays, which influence how the agency of the finished object is manifested. The spirit of the clay is present before the pot is formed. Through the forming, firing, and use of a vessel, this spirit is given its voice through the pot. Each stage built on the spirit is captured in the step before. From the raw material that is the flesh of Grandmother Clay through the cry of the pot when it is broken, potters work with this abstracted agency and shape it into solid form. These arguments have been used to demonstrate the object animacy of special objects (VanPool and Newsome 2012), but the same forces animate quotidian plainwares. The characteristics of materials that solidify in the sherds stay with the object long after it has finished its primary use life.

The case study presented here conveys the impact of the quotidian on materiality studies in archaeology. This is important for several reasons. The first is the omnipresence of domestic ceramics. Compared to imported polychromes, Cerro de Trincheras plainwares were unconsciously used, and the values associated with them were replicated without questioning. The lack of awareness made them powerful objects, precisely because they were normative and not open to challenge (Miller 2005). We see the power of unconscious repetition of daily activity in the *shicra* (bagged construction methods) of Late Archaic Peru (Young-Wolfe, this volume). The impact of the ritual in the domestic is a common theme in Southwestern archaeology (Fowles 2013; Stinson 2010) where the two entities are not seen as dichotomous. Through everyday production in domestic contexts, household vessels perpetuated the values associated with the living clay and the process by which the vessel became animated.

Gruner (this volume) and others (Mills and Ferguson 2008; VanPool and Newsome 2012) use bundling to discuss how ritual paraphernalia is more than a simple sum of its parts. In a similar vein, a pot is more than just a collection of hydrous aluminum phyllosilicates; it is an object enchained by the process of its creation. Cross culturally we see the significance of the tactile transformation from clay to finished ceramic (Hodder 2011; Vitelli 1999; Wengrow 1998). While the clay is animate, it requires the potter to complete the process; in this way the agency of the pot is always secondary to that of its maker (Gell 1998).

Like other chapters in this volume, this case study represents the operationalization of materiality theory. Many authors (Gell 1998; Boivin 2005; Chapman 2000) acknowledge secondary agency in the everyday, yet case studies from the Southwest tend to focus on the extraordinary (Walker and Burt 2009; VanPool and Newsome 2012) and ritual (Mills and Ferguson 2008). Like Coelho (this volume) and Fullen (this volume), my case study addresses how everyday objects gain their power from their omnipresence. Additionally, the power of these broken objects continues into the present in new ways, manifesting itself in debates over national patrimony.

Acknowledgments

I would like to express my sincere gratitude to the many people who assisted me in improving this chapter. Randy McGuire (Binghamton University – SUNY) and María Elisa Villalpando (Centro INAH Sonora) were very generous with their time and knowledge, sharing the archaeology of Cerro de Trincheras and ceramic production in northern Sonora. In addition to the copious editorial work, Ruth Van Dyke's mentorship during this project was invaluable. Thanks also to everyone who read drafts and made suggestions along the way.

References Cited

Appadurai, Arjun. 1986. Introduction: Commodities and the Politics of Value. In *The Social Life of Things*, edited by Arjun Appadurai, pp. 1–63. Cambridge University Press, Cambridge.

Arnold, Dean E. 1991. Can We Go Beyond Cautionary Tales? In *The Ceramic Legacy of Anna O. Shepard*, edited by Ronald L. Bishop and Fred W. Lange, pp. 321–345. University Press of Colorado, Niwot, Colorado.

Boivin, Nicole. 2008. *Material Cultures, Material Minds: The Impact of Things on Human Thoughts, Society and Evolution*. Cambridge University Press, Cambridge.

———. 2005. Geoarchaeology and the Goddess Laksmi: Rajasyjani Insights into Geological Methods and Prehistoric Soil Use. In *Soils, Stones, and Symbols: Cultural Perceptions of the Mineral World*, edited by Nicole Boivin and Mary Ann Owoc, pp.165–186. UCL Press, London.

Brody, J. J., and Rina Swentzell. 1996. *To Touch the Past: The Painted Pottery of Mimbres People*. Hudson Hill Press, New York.

Bunzel, Ruth L. 1972 [1929]. *The Pueblo Potter: A Study of Creative Imagination in Primitive Art*. Dover Publications Inc., New York.

Chapman, John. 2000. *Fragmentation in Archaeology: People, Places and Broken Objects in the Prehistory of South Eastern Europe*. Routledge, London.

Charley, Karen K., and Lea S. McChesney. 2007. Form and Meaning in Indigenous Aesthetics: A Hopi Pottery Perspective. *American Indian Art Magazine* 32(4):84–91.

Colwell-Chanthaphonh, Chip, and T. J. Ferguson. 2006. Memory Pieces and Footprints: Multivocality and the Meanings of Ancient Times and Ancestral Places Among the Zuni and Hopi. *American Anthropologist* 108(1):148–162.

Crown, Patricia L. 1995. The Production of Salado Polychrome in the American Southwest. In *Ceramic Production in the American Southwest*, edited by Barbara J. Mills and Patricia L. Crown, pp. 142–166.

Cushing, Frank Hamilton. 1886. A Study of Pueblo Pottery as Illustrative of Zuñi Culture Growth. *Fourth Annual Report of the Bureau of Ethnology, 1882–83*, pp. 467–522. Smithsonian Institution, Government Printing Office, Washington.

Fontana, Bernard. 1962. *Papago Indian Pottery*. University of Washington Press, Seattle.

Fowles, Severin. 2013. *An Archaeology of Doings*. School of Advanced Research Press, Santa Fe.

Gallaga, Emiliano Murrieta. 2011. Tepalcates Trinchereños: The Ceramic Analysis from Cerro de Trincheras. In *Excavations at Cerro de Trincheras, Sonora, Mexico*, edited by Randall McGuire and María Elisa Villalpando, pp. 93–110. Archaeological Series 204. Arizona State Museum, Tucson.

Gell, Alfred. 1998. *Art and Agency: An Anthropological Theory*. Oxford University Press, Oxford.

Hays-Gilpin, Kelley. 2008. Life's Pathways: Geographic Metaphors in Ancestral Puebloan Material Culture. In *Archaeology Without Borders: Contact, Commerce and Change in the US Southwest and Northwest Mexico*, edited by Laurie Webster and Maxine E. McBrinn, pp. 257–270. University Press of Colorado, Boulder.

———. 2011. Crafting Hopi Identities at the Museum of Northern Arizona. In *Unpacking the Collection: Networks of Material and Social Agency at the Museum*, edited by Sarah Byrne, Anne Clarke, and Rodney Harrison, pp. 185–208. Springer, New York.

Hodder, Ian. 2011. Human-Thing Entanglement: Towards an Integrated Archaeological Perspective. *Journal of the Royal Anthropological Institute* 17(1):154–177.

———. 2012. *Entangled: An Archaeology of the Relationships between Humans and Things*. Wiley-Blackwell, Malden, MA.

Holtorf, Cornelius. 2002. Notes on the Life History of a Pot Sherd. *Journal of Material Culture* 7(1):49–71.

Kopytoff, Igor. 1986. The Cultural Biography of Things: Commodification as Process. In *The Social Life of Things*, edited by Arjun Appadurai, pp. 64–94. Cambridge University Press, Cambridge.

Lechtman, Heather. 1977. Style in Technology: Some Early Thoughts. In *Material Culture: Styles, Organization, and Dynamics of Technology*, edited by H. Lechtman and R. Merrill, pp. 3–20. American Ethnological Society, St. Paul.

Lemmonier, Pierre. 1986. The Study of Material Culture Today: Towards an Anthropology of Technical Systems. *Journal of Anthropological Archaeology* 5:147–186.

Lumholtz, Carl. 1912. *New Trails in Mexico*. Charles Scribner's Sons, New York.

Marshall, Michael P. 1997. The Chacoan Roads—A Cosmological Interpretation. In *Anasazi Architecture and American Design*, edited by Baker H. Morrow and V. B. Price, pp. 62–74. University of New Mexico Press, Albuquerque.

Martínez, María. 1988. *María Martínez: The Indian Pottery of San Ildefonso*. Video Recording. National Park Service.

McGuire, Randall. 2008. *Archaeology as Political Action*. University of California, Berkley.

McGuire, Randall, and María Elisa Villalpando. 2011. Introduction. In *Excavations at Cerro de Trincheras, Sonora, Mexico*, edited by Randall McGuire and María Elisa Villalpando, pp. 1–40. Archaeological Series 204. Arizona State Museum, Tucson.

———. 1993. *An Archaeological Survey of the Altar Valley, Sonora, Mexico*. Archaeological Series 184. Arizona State Museum, Tucson.

Miller, Daniel. 2005. Materiality: An Introduction. In *Materiality*, edited by Daniel Miller, pp. 1–50. Duke University Press, Durham.

Mills, Barbara J., and T. J. Ferguson. 2008. Animate Objects: Shell Trumpets and Ritual Networks in the Greater Southwest. *Journal of Archaeological Method and Theory* 15(4):338–361.

O'Donovan, Maria. 1997. Confronting Archaeological Enigmas: Cerro de Trincheras and Monumentality. Unpublished PhD dissertation, Department of Anthropology, Binghamton University, New York.

Parsons, Elsie Clews. 1996 [1939]. *Pueblo Indian Religion Volumes 1 and 2*. Bison Books, University of Nebraska Press, London.

Pauketat, Timothy R. 2013. Bundles in/as/of Time. In *Big Histories, Human Lives: Tackling Problems of Scale in Archaeology*, edited by John Robb and Timothy R. Pauketat, pp. 35–56. School of Advanced Research Press, Santa Fe.

Schiffer, Michael B. 1999. A Behavioral Theory of Meaning. In *Pottery and People: A Dynamic Interaction*, edited by James M. Skibo and Gary M. Feinman, pp. 199–218. The University of Utah Press, Salt Lake City.

Stevanovic, Marjana. 1997. The Age of Clay: The Social Dynamics of House Destruction. *Journal of Anthropological Archaeology* 16:334–395.

Stevenson, Matilda Cox. 1904. *The Zuñi Indians: Their Mythology, Esoteric Fraternities and Ceremonies*. Annual Report of the Bureau of American Ethnology, Washington.

Stinson, Susan L. 2010. Gender, Household Ritual and Figurines in the Hohokam Regional System. In *Engendering Households in the Prehistoric Southwest*, edited by Barbara J Roth, pp. 116–135. University of Arizona Press, Tucson.

Thomas, Julian. 1996. *Time, Culture, and Identity: An Interpretive Archaeology*. Routledge, New York.

Van Keuren, Scott. 2006. Decorating Glaze Painted Pottery in East-central Arizona. In *The Social Life of Pots*, edited by J. Habicht-Mauche, D. Huntley, and S. Eckert, pp. 86–104. The University of Arizona Press, Tucson.

VanPool, Christine S., and Elizabeth Newsome. 2012. The Spirit in the Material: A Case Study of Animism in the American Southwest. *American Antiquity* 77(2):243–262.

VanPool, Christine S., and Todd VanPool. 2007. *Signs of the Casas Grandes Shamans*. University of Utah Press, Salt Lake City.

Villalpando, María Elisa. 2011. El Tratamiento Mortuorio. In *Excavations at Cerro de Trincheras, Sonora, Mexico*, edited by Randall McGuire and María Elisa Villalpando, pp. 389–402. Archaeological Series 204. Arizona State Museum, Tucson.

Vitelli, Karen D. 1999. "Looking Up" at Early Ceramics in Greece. In *Pottery and People: A Dynamic Interaction*, edited by James M. Skibo and Gary M. Feinman, pp. 184–198. The University of Utah Press, Salt Lake City.

Viveiros de Castro, Eduardo. 1992. *From the Enemy's Point of View: Humanity and Divinity in an Amazonian Society*. University of Chicago Press, Chicago.

———. 1998. Cosmological Deixis and Amerindian Perspectivism. *Journal of the Royal Anthropological Institute* 4:469–488.

Walker, William, and Chadwick Burt. 2009. New Directions in Late Prehistoric Southwest New Mexico: Animacy and Archaeology. In *Quince: Papers from the 15th Biennial Jornada Mogollon Conference*, edited by Mark Thompson, pp. 67–72. El Paso Museum of Archaeology, El Paso, Texas.

Weismantel, Mary. 2013. Inhuman Eyes: Looking at Chavin de Huantar. In *Relational Archaeologies: Humans/Animals/Things*, edited by Christopher Watts, pp. 21–41. Routledge, London and New York.

Wengrow, David. 1998. The Changing Face of Clay: Continuity and Change in the Transition from Village to Urban Life in the Near East. *Antiquity* 22:83–95.

CHAPTER FIVE

An Empire of Clay

Ceramics and Discipline in the Early Modern Portuguese Empire

Rui Gomes Coelho

Introduction

The ascendance of the modern state, with its ideas of absolute and centralized government, was accompanied by the rise of what Foucault called *biopower*, the set of techniques and discourses organized to control the bodies of the subjects and make them internalize ideal frames of power. Power is negotiated in all spheres of public and private life while new mechanisms are disseminated. New forms of exercising power had to make their way through the recognition of a heterarchical world, building from the interdependence of powers that characterized the *ancien régime*, at a time before the rise of the nation-state and modern capitalist societies (Hespanha 1994). The new forms of power were substantiated in material terms through public buildings, squares, urbanism, military, and trade facilities (Moreira 1998). Some authors recognized the articulation of this monumentalization process with other forms of visible power, namely public ceremonies, ephemeral art, and intangible processes, such as rituals and music (Bebiano 1987).

In their daily lives people dealt with the new forms of power in diverse ways: internalizing, resisting, and finding alternatives. Being human implied being bundled with a material landscape that mirrored people in their daily lives. As Foucault stated, "government is the right disposition of things. . . . I think it is not a matter of opposing things to men but, rather, of showing that what government has to do is not territory but, rather, a sort of complex composed of men and things" (Foucault 2000:208). Important social and political aspects of the reorganization of powers and the rise of the modern state were conveyed through daily interactions with objects.

In this chapter, I focus on the Portuguese Empire through the sixteenth, seventeenth, and eighteenth centuries, and I look specifically at ceramics as *extended minds* (Gell 1998) conveying ideal rules of social presence. I argue that the use of different types of ceramics meant more than technological availability, functionality, or symbolism. Early modern Portuguese people had the ability to produce and consume earthenware or glazed ceramics on a wide scale. Among other uses, they were incorporated into the preparation of food and the storage of aliments and beverages (water or wine for example). Ceramics were also used in construction work, in economic activities, such as fishing, and in religious rituals. Because they were widely used, ceramics became a powerful lexicon shared throughout vast lands.

For Alfred Gell (1992, 1998), objects are invested with values through which they participate and as a result influence social life. Objects can be considered enculturated beings animated by collective ideas and social relations that act upon people or other objects. Gell discusses a familiar example of this process: Christian reliquaries, artifacts that serve as depositories of holy body fragments of Saints, and related materials. These reliquaries, used since the Middle Ages, can assume a human form and are animated through the insertion of a relic. Believers anticipate that the Saint, incorporated in the object, will act according to their expectations (Gell 1998:141–143). In this chapter I develop the idea that we can understand how modernity arose through the production, circulation, and consumption of another kind of object—ceramics. Certain broad ideas animated such objects, which then acted as extended minds throughout the Portuguese Empire.

Between the conquest of Ceuta in North Africa (1415) and the independence of Brazil (1822), the kingdom of Portugal was involved in a process of colonial expansion that by the time of its highest dispersion at the end of the sixteenth century encompassed commercial centers, cities, fortresses, and territories from Japan to South America (Newitt 2005). The structure changed throughout this period from an essentially maritime and commercial power in the sixteenth century to a territorialized empire in the seventeenth and eighteenth centuries as domains like Brazil and Goa were expanded. But what is more significant to our discussion is the ideological shift that occurred in the second quarter of the sixteenth century. From the Messianic ideology that justified the expansionist project during the reign of Manuel I ([1495–1521] i.e., a crusader's mentality that envisioned the creation of a unified Christian empire under his aegis), the Portuguese monarchy moved in a different direction. Contact with China and other powerful Asiatic polities challenged the perspective of a universal empire headed by a Portuguese king. Under John III (1521–1557) the Portuguese Kingdom started to be seen as the reconstruction of the old Roman Empire, this time as a project of "global reconnection" controlled by the Portuguese monarchy, which contained ideological space for non-Christian powers like China (Biedermann 2014). As I discuss below, these transformations are crucially bound up with the roles and uses of porcelain and faience.

If one considers that the entire Portuguese European population only surpassed the three million barrier in the late eighteenth century or early nineteenth century (Machado 1965), it is critical to examine the ways elites, especially those related to the crown, dealt with the integration of such a diverse empire in daily practice. Several archaeologists and historians have documented the complex trade and labor networks that spread ceramics and especially faience, which was produced in three main centers of Vila Nova de Gaia, Coimbra, and Lisbon and ultimately used all over the world (Sebastian 2011). In some cases, these ceramics were also part of minor exportations to the rest of Europe and European colonies (Casimiro 2011); eventually they were related to the Portuguese Jewish diaspora (Baart 1988). But these faience studies lack a more comprehensive investigation into why people chose certain objects instead of others, what meanings they carried, and what roles they played in social relations.

In order to understand this complex relationship between ceramic production and consumption and power construction, I look at a specific archaeological site in the Portuguese town of Setúbal. The ceramics I focus on in this chapter come from the excavation of a building in the neighborhood of Troino in Setúbal and are associated with two different contexts: a late sixteenth-century/early seventeenth-century habitation and a middle eighteenth-century context. This town in the south of Portugal has its roots in a late Bronze Age settlement (eighth and seventh centuries BC) but became an important fishing and commercial port after the Christian conquest in the late Middle Ages (early thirteenth century). The western neighborhood of Troino, specifically, has been associated with fishing activities since then (Soares 2000). Although Setúbal is located in Europe and is less than fifty miles south of Lisbon, this maritime town embodied the idea of an empire that identified itself as the intersection of global networks.

The archaeological work at the site revealed three main construction events. In the late sixteenth century, the fishing community built a round cabin on the beach. The foundations of this cabin were made of stone, while the rest of the structure was probably built of wood and other perishable materials. Archaeologists were able to recover storage and kitchen ceramics, modest tableware, and several fishing-related artifacts (Coelho 2010:117–155). In the mid-seventeenth century, soil and trash were deposited on the spot to create the foundation for a stone building. This moment corresponds to a significant change in the history of Setúbal; the town's main economic activity shifted from fishing and other marine activities to the more lucrative salt-production business. This stimulated urban expansion from what was before a relatively peripheral fishermen's beach. The new building had two storage rooms with floors covered with pebbles plus a possible backyard (Coelho 2010:156–200). The building was partially transformed by the end of the seventeenth century. Some of the walls were reinforced, presumably to allow another expansion of a residential upper floor, while the floors were covered with bricks. One of the previous storage areas was transformed into a tavern with an area for wine storage and a small well used to provide water and cool the room. Glass wine bottles, kitchen and tableware, devotional medals, Dutch pipe fragments, and coins are among the material testimonies of this tavern and the residential upper floor. This building was almost completely destroyed by the Great Earthquake on November 1, 1755, and that is why the context was relatively well preserved (Coelho 2010:201–445).

The excavation at Setúbal provided a good record of the material experiences of its inhabitants over 150 years and revealed a fisherman's cabin built in the late sixteenth century/early seventeenth century and a seventeenth-century building destroyed by the Great Earthquake. The ceramic objects from the two contexts, particularly, can be used to compare economic and social shifts in Setúbal in the early modern period (Table 5.1). In the late sixteenth-/early seventeenth-century fisherman's cabin, most of the ceramics were earthenwares (Figure 5.1). In the earthquake destruction level in the two-story building earthenware objects constituted the majority of the ceramics, although their proportions were less significant when compared with the cabin. Meanwhile, the percentage of faience rose from less than 10 percent to almost 40 percent. Porcelains also became an everyday item in Portuguese urban contexts and are visible in the middle eighteenth-century context (Coelho 2010:229–327).

The clear preference for earthenwares in the late sixteenth-/early seventeenth-century contexts at Setúbal reveals how pottery vessels were

Table 5.1. Percentages of ceramics from two of the contexts found at the Largo António Joaquim Correia site, Setúbal

	Late 1500s/early 1600s context	*Middle 1700s context*
Earthenware	86.78	43
Faience	9.43	37.40
Lead Glazed	3.77	16.69
Porcelain	—	1.21
Others	—	1.67

entangled with early modern notions of morality that included humility and gender identity. This did not change entirely during the early modern period. The material remains of the two-story house destroyed in 1755 reveal that almost half of the ceramics were earthenwares, but in the meantime faience and porcelain emerged as important items.

The definition of moralities is closely related to the construction of certain forms of sensorial experience. People in the early modern times understood clay in terms of taste and haptics and how different types of ceramics played distinct roles in sensorial perception. The rise in the consumption of glazed vessels, especially faience and porcelain, brought changes to a domestic world that was essentially made of earthenware. Nonglazed objects were supposed to participate in complex interactions with other elements in the preparation of meals, or recipes, but they gradually disappeared with the rise of odorless and tasteless artifacts. Therefore, the transition from nonglazed clay ceramics to tin-glazed products during the seventeenth and eighteenth centuries may be perceived as part of the discipline of taste and olfaction.

But the changes porcelain and faience introduced can also be regarded in a different way that is intimately tied to the definition of particular moralities and sensations. The rise of faience consumption in the early Portuguese Empire is closely related to the establishment of trade networks in East Asia, the arrival of Chinese porcelain to Europe on a large scale, and the definition of distinctive Portuguese decorative models in faience production centers, especially the well-known blue and white patterns that could be found all over the empire. The linkage between Chinese porcelain and Portuguese faience is symmetrical to the revelation of Chinese society and politics to the Portuguese elites at a crucial moment of the empire's formation.

Figure 5.1. Clay mugs found in a late sixteenth-century fishermen's cabin in Setúbal. Photo by Rui Gomes Coelho, courtesy of the Museum of Archaeology and Ethnography of the District of Setúbal, Portugal.

The use of clay vessels in the construction of moralities and sensorial paradigms and the definition of novel political modes of existence for some types of ceramics cannot be dissociated from the anonymous labor of the potters, as activities of the potters were entangled with broader processes of power negotiation.

Ceramics and Morality

Late seventeenth-century and early eighteenth-century texts provide a wealth of information about attitudes toward ceramic objects. The body, assumed to be fragile and ephemeral, was often compared with clay. The idea of Adam as made of dust (Genesis 2:7)—often interpreted as of clay—has long been present in the Roman Catholic imaginary, and it represents the dichotomy of the celestial vs. mundane and the primordial bundling between humanity and the material world. In the context of the Counter-Reformation, clay and clay-made objects became a metaphor for the human condition, as well as a medium to disseminate proper conduct.

Examples abound in early modern literature. A 1735 text, for instance, clearly associates the body with a clay bowl, while the soul is represented as a jewel inside that object (Jesus 1735:192–193). And, because clay is fragile and ephemeral, it is also a symbol of humility. In a text attributed to Father António Vieira, the author berates the king's officers for using silver objects when they should be contented with "faience from Lisbon"; officers should have recognized that in the conservation of wines, *talhas* (large clay pots) are better than golden jars. Clay was seen as protector of virtues (Vieira 1744 [1652]:250–251, 265). In 1563, Friar Bartolomeu dos Mártires was invited to dinner with the Pope in Rome. During the dinner, while regarding the opulent silver serving sets in the room, Friar Bartolomeu recommended to the Pope the use of porcelains from China, since they were equivalently sophisticated but made of clay (Sousa 1921 [1619]:155–159).

These ideas of fragility, temporality, and virtue also extended to women. For example, an anonymous author stated that "women are like clay—they only break with a heavy fall" (Anonymous 1762 [1760]:27–28). One author described two men's dispute for a women as "an issue concerning the mug of a certain Lady," which led them to take up the steel—the swords (Morais 1761:432). In satirical poetry, comparisons with pottery caricaturize feminine figures; the lips of a dancer or those of the Fame are exaggeratedly described as like those of a big bowl (Académico 1762:162; Freire 1731:34–35). Clay is invested with the moral and physical limits of women through the description of their status or the definition of a frontier of the ridiculous and the socially unacceptable. Such a frontier is also defined regarding the position of the man. As one popular maxim stated, men belong "at the plaza, and women in the house" (Bluteau 1712–1728:5, 546).

The early modern period is characterized by the great power of the Church, economically and politically. Elites often sent second children to convents and monasteries to avoid the high cost of dowries (Faria 1791 [1625]:61–65). Many people living in convents and monasteries were there for social reasons, so they tried to minimize their removal from the secular world through the materialization of their social condition inside the cells. Moral treatises of the 1600s and early 1700s describe the immoral environment of Portuguese nuns as surrounded by domestic and exotic animals, mirrors, highly elaborated laces, paintings, golden and wooden carved works, closets, figurines in alabaster or plaster, handkerchiefs, and pillows and bed sheets made of luxurious fabrics from all over the world. One could find Chinese porcelains, glass artifacts, painted faience, and gold and silver daily objects, such as jars or forks. Friar Manuel de São Luís (1731:432) wrote that nuns' cells were "decorated as if they were churches, with many superfluous things." A major example of this was the apartments of Paula Teresa da Silva e Almeida—King John V's favorite and Mother Superior of the St. Denis monastery at Odivelas (1701–1768); her private area inside the monastery was compared to an aristocratic palace (Anonymous ca. 1750). Religious women were known for preparing candies, bouquets of flowers, and all sorts of refined handicrafts to offer their relatives, friends, and lovers (Bernardes 1699:251–252; Velho 1730:120). Through their complex use of light and shade, still life painters such as Josefa d'Óbidos superbly conveyed the role of objects in the struggle for the morality of an entire society (Hatherly 1993).

The alternative materialities, beacons of a moral life, were clearly defined. Father Manuel Velho, for example, argued that a nun's mug should be made of clay. Her dishes should be of the rudest national faience because "Genoa and India are, and must be, far from a nun." She also should have a tinplate lamp, and her bed should be a simple mat made of wood or cork (1730:313). Christ's wives should engage with the manufacture of devotional objects, bouquets for the Church, or the sewing of their own clothes (Velho 1730).

These texts clearly illustrate that the material world mediated early modern social life and morality. Moreover, humans and materials were bundled through the definition and clarification of such mediation. Clay conveys fragility, ephemerality, humility, and virtue. Clay is also associated with the image of the ideal woman and, therefore, the domestic space and works as an "extension" of those values in the Gellian sense. These connections help us to understand a preference for earthenware ceramics in early modern contexts, such as the building at Setúbal. These vessels represent a connection between women and domestic space, and women and clay. But why did the earlier occupants prefer nonglazed to glazed ceramic containers? Part of the answer lies in early modern concepts about materiality and the senses.

Ceramics and the Senses

Biopower politics include the disciplining of the senses. In the late eighteenth and nineteenth centuries, olfactory landscapes changed completely with a set of practices organized by savants (such as physicians) to systematize and categorize types of air and to define those that could be considered healthy (Corbin 1982). Olfaction became a key instrument in the perception of air, and it carried the attribution of social significance to smells. Something similar probably happened with gustatory and haptic senses. The idea that senses are socially and historically constructed may lead us to understand better certain material choices and their longevity in domestic contexts.

The importance of having tasteless and odorless objects, especially those manipulated in the kitchen and used to consume aliments, is relatively recent in the Western world. Documentary and archaeological evidence suggests that during the early modern period, objects were not chosen according to modern concepts of prestige or hygiene. Rather, they were viewed organically as interactive elements prepared for being combined with multiple substances and in diverse environments. According to the theory of humoralism, the health of the human being is defined by the equilibrium of four fluids, each of them associated with the four elements and temperaments. This system of thinking is essentially behind the ways people dealt with material culture in the ancien régime—a permanent interaction of things, elements, and fluids with the human body. Some European medieval authors reveal this relationship through comparisons and connections between the human body and the use of objects (Alexandre-Bidon 2005:147–154). The doctor Francisco da Fonseca Henriques (1731:11, 21), for example, used the preparation of bread and its fermentation as a parallel of digestion. In recipes, the reaction of a pot to the ingredients was considered important. In the case of a remedy that included the blood of a hare, instructions said to "take a living hare during May, behead it, and drain its blood into a new nonglazed bowl in order to soak up the serum" (Semedo 1697:388). People were aware of chemical interactions, and that is why some doctors did not recommend the use of metal ware to boil water (Semedo 1697:571).

In early modern Portugal, therefore, certain kinds of objects were purposely made to have specific tastes and smells. Ceramic vessels from Lisbon, Montemor-o-Novo, and especially Estremoz, in the southern part of the kingdom, were highly praised for their taste and the way they kept water fresh:

> It [mug] exceeds the beauty of crystal. Even if not by virtue of ethereality, then by the taste it lends to the water one drinks through them. It also pleases the sense of smell thanks to its scent; it is aromatic without any artifice. These mugs, red and beautifully shaped, are pleasant to our eyes and influence most of our external senses, including touch. The clay clings to our lips so tenaciously that if the mug is small, our lips will hold it by the brim. (Henriques 1726:207–208)

This predilection for earthenware caused surprise to foreign visitors. The English subject Thomas Cox visited Portugal in the early 1700s and commented on the Portuguese daily habit of drinking through "stinky" clay mugs. Cox also observed people eating clay objects, especially women (Cox and Macro 2007 [1701]:51 verso). The ingestion of clay was known in Europe as beneficial for medicinal purposes. João Curvo Semedo (1697:660) recommended the storage of infusions in Estremoz mugs and regarded their clay properties as beneficial to cardiac patients. In 1726, the king's doctor described the "Armenian cake"—a ground clay preparation—in his treaty (Henriques 1726:208–209); it was something used during exorcism procedures (Souza 2003:111). But such consumption was also seen as a problem, and this is why Luís Gomes Ferreira (1735:158), a doctor working in the southeast of Brazil, published a remedy recipe to help clay-addicted people. It required a person to urinate in a clay artifact, let the urine dry, and then turn the object into powder and prepare a drink with it. Another suggestion was to drink water with burial dust in a clay object.

The desire for glazed and nonglazed ceramics was so high, even among the upper classes, that travelers felt the need to explain it to their foreign audiences. Some pointed out the "great abundance" of earthenware plates in the aristocratic houses (Cox and Macro 2007 [1701]:48 verso). Venturini, the Italian secretary of Pope Pius V, described with surprise in 1571 the fact that the king Sebastian I always had a clay mug on the table among his silver serving set, which he used to drink water (Vasconcellos 1957 [1921]:13).

The description of such practices was a part of a larger process of domestication of the senses taking place in the society. By integrating this knowledge into written forms, its formal control was directed by certain elites who ultimately became able to reproduce it within their social values. And with such appropriation, elites became able to mobilize and gradually change the senses of the society.

The two-story house destroyed in Setúbal in the Great Earthquake can be regarded as a microcosm of these changes. One of the compartments in the ground level functioned as a tavern, and the vestiges of two large clay pots probably destined to store wine have been found in one of the extremities of the compartment. These talhas were used to store liquids, including wine, in wine cellars and taverns all over the south of Portugal, a tradition from the Roman times that gradually fell out of fashion. In the late nineteenth century, people still noted the repugnance felt by natives of the southern region of Alentejo for drinking wine kept in wooden barrels (Lapa 1868:298).

Also, glass—mostly from bottles—constituted over 40 percent of the objects recovered from the house in Setúbal. In the fisherman's cabin that existed in the same area around 150 years before, there was no glass at all (Coelho 2010:216). All that glass, although in a very fragmentary state, gives the impression that people used more and more glass throughout the eighteenth century, which corresponded to a growing preference for vessels without smell or taste.

Power, Blue, and White

Porcelain and faience are key categories of objects in this process. They are not just participants in the transformation of the senses as I discussed above, but they are also secondary agents of power strategies in a Gellian perspective (1998). Faience became a crucial vehicle for the distribution of ideas of empire.

Although very limited quantities of porcelain reached Western Europe through the Silk Road before the end of the fifteenth century, the arrival of the Portuguese in India in 1498 and the consequent opening of a sea route to Asia allowed a regular and intense trade of Chinese ceramics. Blue and white porcelains from the Ming Dynasty (1368–1644) are represented in European arts for the first time in 1514 with Bellini and Titian's *The Feast of the Arts*. This was only about one year after the arrival of the European sailors in China, and it confirms the immediate impact of such materials. This trade only increased significantly during the 1520s when the Portuguese crown requested that cargo ships from India carry porcelains up to one-third of their capacity. From the very beginning, the Portuguese king was interested in such ceramics and ordered the production of some samples with the royal symbols, namely the armillary sphere (Monteiro 1994:19). Beyond capitalist values of material novelty, there are two main interconnected reasons for the success of Ming ceramics in the Portuguese court, which in a certain way may be extended to the rest of the Western world.

First, the circulation of such ceramics seems to correspond to a flux between the idea of empire as directed by a ruling monarchy that had exclusive access to maritime routes and its subjects' daily lives. Color played an important role here. Colors were determinant in the construction of power over the ancien régime all over Europe and were regarded according to specific principles and morals. For instance, sumptuary laws considering economic and social statuses, as well as ethnic or religious identities, regulated the usage of colors. Various thinkers, including theologians, jurists, and others discussed colors' values and meanings in terms of early scientific discussions on the physical nature of color. In 1683, the early modern jurist Hermann Wissmann assumed in his dissertation about law and colors that the meaning of colors were inscribed in all things as elements of a natural order that should be visible to everyone (see Hespanha 2006).

The dominant tones in porcelain were blue and white (even if the white was not a pigment), and the blue color in the late 1400s through the 1500s was entering a new process of signification. During the late Middle Ages, blue became associated with royal power and Marian devotion. With the Reformation, it also became a moral color; for the Protestants, blue was a sober color and engaged with black in the definition of a true "chromoclasm," analogous to the reform-inspired "iconoclasm" (Pastoureau 2001:100). If the role of sober colors, previously absent in Catholic liturgy, are understandable in the Reformation process and help to explain the relative success of blue and white ceramics in Netherlands, for example, in the Counter-Reformation ambiance of southern Europe, things might have been very different. As Pastoureau recognized, the Church is an image of heaven on earth, and all the colors are invited to participate in such magnificence, especially gold (Pastoureau 2001:104–108). The combination of gold and blue, but also blue and violet, were very popular in Portugal in the early eighteenth century and the second half of the seventeenth century. They were represented in architectonic tiles and in faience and can be understood as a connection between blue (royal and Marian color and imperial color due to the contact with the Chinese) and the Counter-Reformation's gold.

This is not a coincidence considering that the modern state was in gestation, and new forms of power negotiation were being explored. As Graeber (2004:70) suggests for the rise of nation-states, "Western elites were trying to model themselves on China, the only state in existence at the time which actually seemed to conform to their ideal of a uniform population, who in Confucian terms were the source of sovereignty, creators of a vernacular literature, subject to a uniform code of laws, administered by bureaucrats chosen by merit, [and] trained in that vernacular literature." Such a process probably started much before the nineteenth century. As I mentioned above, the second quarter of the sixteenth century was marked by a significant shift in the way the Portuguese elites constructed a narrative to legitimate the empire. The contacts with China challenged the idea that a universal Christian empire led by the Portuguese king would ever become a reality. However, those contacts also stimulated a growing interest in Chinese society and political organization along with the revisitation of classical traditions. The Portuguese imperial project became the intersection of two symmetrical powers: the Chinese and the Roman empires (Biedermann 2014).

Although we do not know the relative costs of porcelain objects, it is certain that they were expensive. The archaeological record typically contains small bowls and dishes, usually with more or less simple decorative patterns. In the commercial and domestic context destroyed in 1755 in Setúbal, porcelain accounted for only 1.2 percent of the ceramics; although the owners of this place kept a dish from the time of Emperor Jiajing ([1522–1567] Figure 5.2). We can compare this context with earlier data to form a better idea. As mentioned before, the building was built on trash deposits in the seventeenth century. The workers probably brought dirt and trash from different parts of the town, so we cannot associate the archaeological materials with any specific original context. However, we know that most of the ceramics predate the middle seventeenth century. Porcelains were not very common at all (Coelho 2010:174, 216, 297–298). Thus, we may assume that the most significant artifacts were destined for elite consumption, and those are the samples that have made their way into museums and private collections (Monteiro 1994).

Faience, however, may have worked as an alternative to porcelain and allowed a wider social distribution. Tin-glazed ceramics started to be produced in Europe in the late Middle Ages, but the techniques probably arrived in Portugal during the sixteenth century with the migration of Flemish technicians. They were established in Lisbon by at least 1565, and in the early seventeenth century it is possible to find references to the production of ceramics in the "Venetian" way, which is certainly faience. Concerning the trade of blue pigments to be applied to ceramics, the first reference appeared in the middle of the seventeenth century, but it is certain that such a technique was in use for a long time (Sebastian 2011:118–119, 394–395). The production of such ceramics soon spread from Lisbon to two other main centers: Vila Nova de Gaia and Coimbra, both very important in ceramic distribution in the northern and central part of the kingdom. All of them—Coimbra in much smaller quantities—were exported overseas and especially to the colonies (Sebastian 2011:167, 194).

Figure 5.2. (opposite page) Large Chinese porcelain plate found in the Great Earthquake destruction level of a house in Setúbal. It is a good example of a blue and white artifact produced during the Ming Dynasty when Jiajing was emperor ([1522–1567] Coelho 2010:412). Drawing by author.

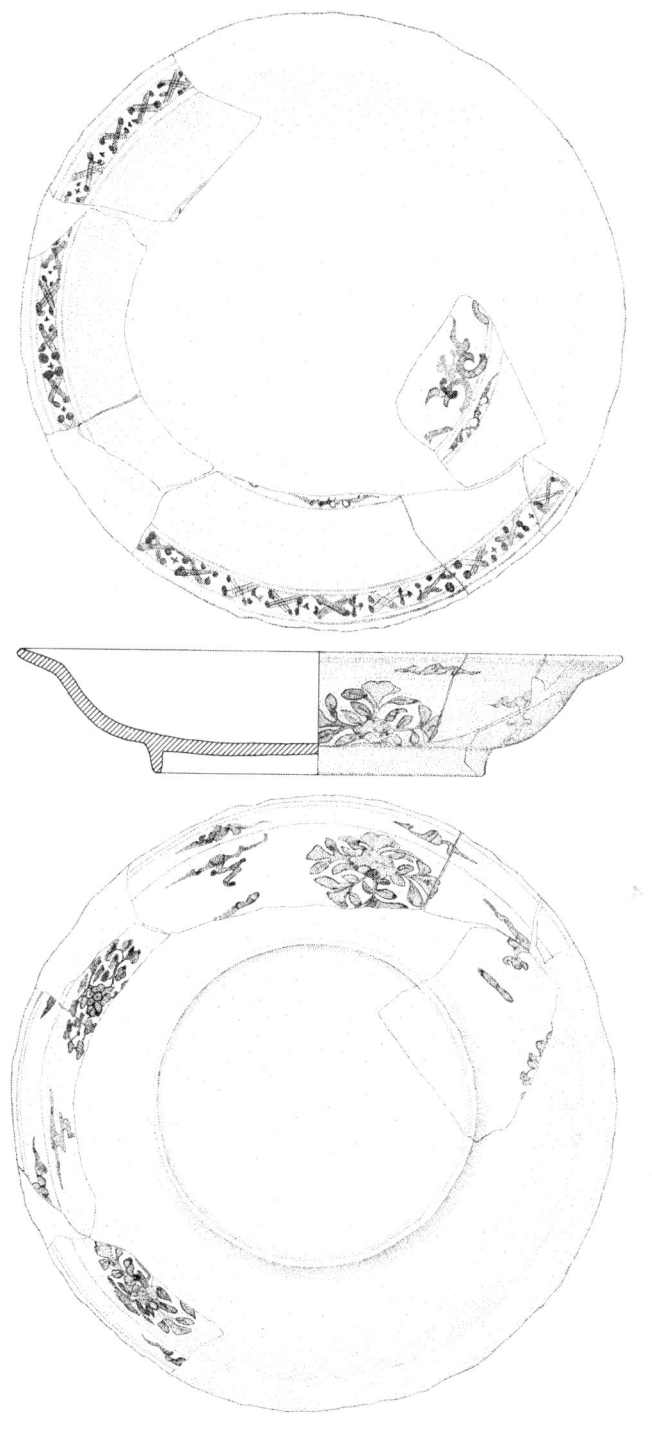

5 cm

Azulejos (tin-glazed painted tiles) accompanied the growing production of faience vessels. However, azulejos were mostly for architectonic use and had a greater public significance (Simões 1997).

The first half of the 1600s is considered to be the apogee of Portuguese faience in terms of technical quality and aesthetic elaboration. Even if they were not the only source of inspiration for its decoration, Ming porcelains affirmed a new aesthetic approach to tin-glazed ceramics; through the years, Chinese decorations and patterns were imitated, combined with European iconography, and schematized. As happened with porcelain, highly elaborated artifacts remained inaccessible to the general population. However, as the archaeological record shows, those wealthy ceramics may have been seen as paradigmatic ones, and the increase of faience consumption was explosive between the sixteenth and the middle of the eighteenth centuries. Faience could be found everywhere—from a mission in Amazonia to an urban center in Portugal. In the house destroyed in Setúbal in 1755, faience accounts for 37.4 percent of the ceramics—mostly tableware ([Figure 5.3] Coelho 2010:216, 221). When in the late sixteenth century the area was just a beach, and was occupied by a fishermen's cabin, the amount of faience was only 9.4 percent of the ceramics, and none of the objects had a Chinese pattern (Coelho 2010:131).

It seems that these kinds of objects became part of a broad process of integration of a vast territory through the discipline of daily lives; while the elites had access to Chinese porcelains and to the most elaborated European faiences, the general population became part of a new material frame. Shapes, decorations, and colors all became standardized, and they appealed to a certain social order; they were a notion of empire under distribution in the Gellian sense. The development of faience workshops in a new area of Lisbon, closer to the river and in opposition to the old Moorish potteries closer to the medieval castle (Sebastian 2011:512), may not have been a coincidence. We can find there a parallel with the decision of the king in the late 1400s to move from his residence in the castle to a new palace next to the harbor (Senos 2002).

Enchanting Potters

Pottery production occupies in itself a central role in the use of ceramics as a vehicle embedded with moralities and political values. While discussing art as technical system, Alfred Gell argued that artistic production involves social consequences as an ultimate goal. Thus, the power of art objects

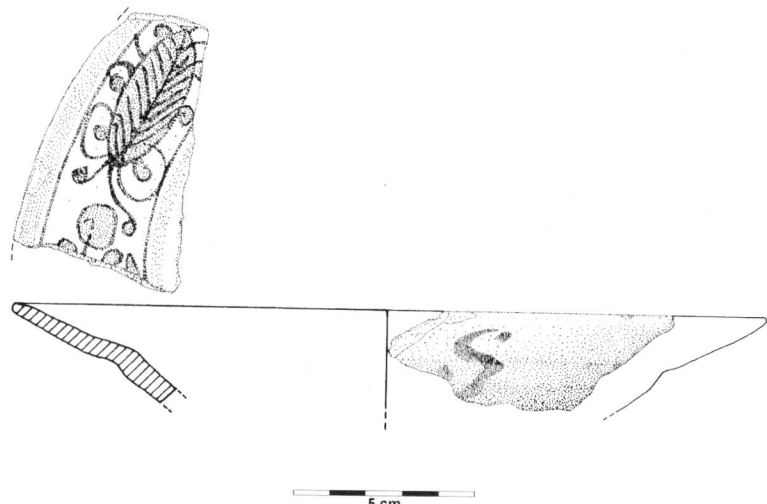

Figure 5.3. Faience plate found in the Great Earthquake destruction level of a house in Setúbal. It shows Chinese-inspired decorations (blue, purple, and white) and was probably made in Lisbon between 1625 and 1700 (Coelho 2010:388). Drawing by author.

is directly related to the technical processes they incarnate. Gell further explains this idea with the categories of craftsmanship as "technology of enchantment" and its results as "enchantment of technology." In other words, the objects engage with the social world (i.e., they "enchant") as the result of a technical process that is regarded as complex, specialized, and quasi magical (Gell 1992:44–47). The consequences of this process are important for understanding the production of ceramics as a whole in the context of the early modern Portuguese Empire. For Gell, the objects become a physical mediator between two beings, establishing and enhancing social relations. Thus, the "magical power" of an artist working for the king is symmetrical to the effect the monarch expects to exert among his subjects (Gell 1992:52).

An intricate process can be recognized in the social role potters assumed in early modern times, as they mediated between a monarchy under construction and a myriad of diverse subjects that the elites intended to integrate within the same sphere. As discussed below, the formal regulation of pottery making and the public engagement of the potters with the material agendas of the elites resulted in the creation of a particular material world that is specific to this period. The explanation of how this came to be is crucial for understanding the context of the archaeological site in Setúbal.

In early modern Europe, the manufacture of ceramics gradually became a highly specialized activity. As I discussed before, the emergence of faience production on a wide scale corresponded to a growing technical sophistication among kiln builders, potters, and pottery painters. The same could be said for the increasing production of earthenwares or other glazed ceramics, especially if we take into consideration the expansion of colonial ventures.

From the late Middle Ages through early modern times, ceramic production became increasingly normalized through detailed regulations, especially in larger cities, such as Lisbon (1572), Coimbra (1623), and even small urban centers like Aveiro (1727). These regulations were related to the professionalization of potters by not only enumerating the objects they should have been able to make and the required techniques, but also by how authorized examiners should inspect potters' works. Besides that, there were regulations concerning the normalization of prices of pottery and related services, such as its transportation. These regulations were more common than those of the craftsmen and were found throughout the ancien régime. This was the case at regional centers, such as Lamego (1530), border towns like Elvas (1632), important colonial ports like Funchal in the Madeira Island (1587), and at Angra in the Azores (1788) (Fernandes 2013).

Although local authorities were responsible for most of the regulations, the wide spread of these written rules was, to a certain extent, a tool for the standardization of production and consumption practices and terminology. In certain moments and regions, concerns for the standardization of practices became more apparent. It happened in the north of Brazil over the government of the Marquis of Pombal as secretary of state (1750–1777). Determined to transform the Amazonian region into an enlightened model of a colony, the Portuguese crown stimulated the establishment of potteries in indigenous villages, which seemed to have led to the wide spread of European-type decorations on domestic paraphernalia (Domingues 2000; Coelho 2009:178–179).

As noted, the standardization of professional practices was in gestation at least since the late Middle Ages, and it contributed to underline the special status of the artisans as crafters of political powers. In 1619, when King Philip III of Castille and II of Portugal visited Lisbon, several professional organizations and foreign communities erected a series of ephemeral arches throughout the city to honor the sovereign or assert what they considered to be their contributions to the construction of the monarchy. Among businessmen, tailors, shoemakers, or painters were also the potters, but unfortunately only a written representation of their ephemeral arch survived the centuries (Lavanha 1622:29–30). While passing by the arch, the king would have seen the potter's patrons, the martyrs, and sister Saints Justa and Rufina handling clay vessels along with a tower constructed between the figures. In the arch one would have also observed the allegoric painting of *Nature* crowned with flowers, handling in one of its hands a red clay vessel and in the other a half buried man that represented clay itself. There was another painting, an allegoric depiction of *Art*. Its right hand was on a potter's wheel, and its left was handling a faience object, described by the chronicler João Baptista Lavanha as "a porcelain vessel made in Lisbon counterfeit from China" (Lavanha 1622:30). Besides the court few people were certainly able to read the poems that accompanied the sculptures and allegories, but they were nevertheless significant. For instance, the potters displayed a text on the tower that suggested that although the tower had its foundations made of clay, the fact that it was built to honor the king was enough to strengthen the tower. A quartet was associated with the "art" of handling a faience vessel:

> Here, excelsior and sovereign monarch
> The pioneering Art offers you
> That which is made in the Lusitanian kingdom
> And before expensively sold to us by China.
> (Lavanha 1622:30)

The potters and those who read this text certainly knew that Chinese porcelain differed significantly from European faience. However, what was most important here was the idea of capturing the values associated with the porcelain. Thus, while the potters were allegorically emulating their technical skills from the crown's power, they were also magically contributing to the reinforcement of its authority. The images of Portuguese ships in the harbor on Chinese porcelain and foreign ships on Portuguese ceramics might have been one of the most formidable images in the arch, and undeniably it conveyed the message of power construction (Lavanha 1622:30).

The potters were crafting, in their own way, the construction of the monarchy through ceramics and disseminating it throughout the material world. And they did so because they were invested in the monarchy's authority, either through the discipline that bounded their activities or through the recognition of their contribution and the elite consumption of their products. The producers of faience and painted tiles certainly achieved special esteem. In the Gellian sense (1992), pottery craftsmanship literally became a technology of enchantment.

Conclusion

Throughout this chapter I have discussed how ceramics were used during the early modern times to convey moralities, notions of political power, and to discipline the senses. Amidst the extremely complex process that is the rise of modernity, the definition of material narratives that could be extended among the maximum number of subjects was crucial for the negotiation of powers and for the construction of the modern state. Ceramics were part of it, and the analysis of their production and consumption offers a different portrait of this process.

Alfred Gell's notion of extended mind enables us to go beyond issues of representation and production (Gell 1998:221–258). While analyzing Marquesan art, Gell understood artworks as constitutive parts of a *distributed object* in time and space (Gell 1998:221). In spite of being separated and disseminated, Marquesan objects were all interrelated through social networks, and as a whole they formed a Marquesan mind. The metaphor Alfred Gell uses to exemplify his concept is a set of China tea and dinnerware, a group of artifacts that makes sense as an assemblage. Gell recognized that Marquesan art could not be understood as the result of a centralized and deliberated organization. A similar process was happening with ceramics in the early modern Portuguese Empire. Ceramics worked as an extended mind; they carried notions of appropriate social conduct and contributed to the discipline of the population throughout the empire.

We have already seen how clay-made artifacts, and especially earthenwares, were connected with ideas of fragility, ephemerality, humility, and virtue. By writing discourses on clay, the elites normalized their moral associations and helped spread them through the work of the potters. But this discipline can be also understood in the ways the senses were constructed. We could see, for example, how people gradually preferred tasteless and odorless artifacts even if "stinky" earthenwares were still very much part of quotidian practices. Those odorless and tasteless objects, such as faience and porcelain, were imposed on domestic lives in different ways. New notions of hygiene probably contributed to it, and the rise of modern printing and the appropriation of such ideas and relationships through a written discourse during the 1600s and 1700s helped the domestication of sensorial experiences by the elites close to the monarchical power. Then, those circles could manage to convey more easily their social values and practices, especially regarding the body and the senses.

However, the arrival of the Portuguese in China in the 1520s and the subsequent popularity of porcelain in Europe was connected to the rise of faience production and consumption throughout the Portuguese Empire. As we can observe in the modest context of Setúbal, the seventeenth and the eighteenth centuries were marked by the rise and spread of faience consumption that seems to correspond to the distribution (Gell 1998) of an idea of empire. Those tin-glazed and decorated plates, bowls, or jars were not just symbols of a successful monarchy that controlled territories, people, and fluxes of richness all over the world, they were agents of the empire itself, and their dissemination conveyed relationships and social roles.

The eighteenth century is also important in a different way; the humoralist vision of the world was in decline, and with such transformations new conceptions of hygiene emerged. Faience—and glass—had a central role in this process because these materials represented the object without taste and without smell (Alexandre-Bidon 2005:167–170). It was also the century when ceramics became more and more fragile together with the faster production and consumption that characterizes capitalist relations (Duhart 2001:245).

In 1761, the intellectual nobleman Francisco Xavier d'Oliveira was burned in effigy in one of the last inquisitorial public executions in Lisbon. He had been living in England since 1744, and in 1746 he abjured Roman Catholicism. Although he died peacefully in his house in 1783, for Portuguese society as a whole this public mock execution carried a deep significance and ultimately meant death itself. As Tavares recognized, in a context where people lived continually between symbols and things, such an execution was terrifying. It announced the end itself through a state of melancholy and slow death by sadness and shame (Tavares 2005:168).

However, the exertion of social power and its materialization were never as clear as the individual execution of an effigy person. The integration of a vast and diverse number of subjects all over the world under common morals, sensory perceptions, and political notions occurred in a much more discrete way. Amidst the remains of a tavern or residence in a neighborhood of Setúbal, we could imagine bowls of faience and potsherds of earthenware as little effigies of ideas that were extended, conveyed, and distributed throughout the empire.

Acknowledgments

This chapter is partially based on research I have done at the Museum of Archaeology and Ethnography of the District of Setúbal, Portugal. My thanks to its director Joaquina Soares and to the archaeologists who conducted excavations at the site I studied: Carlos Tavares da Silva, Antónia Coelho-Soares, and Susana Duarte. I also wish to thank Rosa Varela Gomes, who advised my MA thesis. The text benefited from the insights of Ana Rita Trindade, António Manuel Hespanha, Hande Sarikuzu, Mark Hauser, Patrícia Melo, and the anonymous reviewers of the University of Arizona Press. Zoltán Biedermann and Isabel Maria Fernandes kindly provided me drafts of forthcoming works that contributed to this discussion. I am especially grateful to Ruth Van Dyke and the stimulating group of Binghamton University – SUNY friends and colleagues participating in this volume, which is the result of several months of discussions and great food exchanges.

References Cited

Académico. 1762. Cantava huma Dama, e Fabio sem a ver se Enamorou só por ouvi-la. Assumpto Académico. In *Eccos que o clarim da fama dá: Postilhão de Apollo, montado no Pegazo, girando o Universo, para divulgar ao orbe literario as peregrinas flores da poezia portuguesa, com que vistosamente se esmaltão os jardins das Musas do Parnazo*, Vol. 2, pp. 127–135. Oficina de Francisco Borges de Sousa, Lisbon.

Alexandre-Bidon, Danièle. 2005. *Une archéologie du goût. Céramique et consommation (Moyen Âge—Temps modernes)*. Picard, Paris.

Anonymous. ca. 1750. Description of Mother Paula Apartments in the Convento of Odivelas. Cód. 68//10. National Library, Lisbon.

Anonymous [F. J. C. D. S. R. B. H]. 1762 [1760]. Conferencia IV. In *Academia dos humildes, e ignorantes. Dialogo entre hum theologo, hum filosofo, hum ermitão, e hum soldado, No sitio de Nossa Senhora da Consolação. Obra utilíssima Para todas as pessoas ecclesiasticas, e seculares, que não tem livrarias suas, nem tempo para se aproveitarem das publicas*. III:25–37. Oficina de Inácio Nogueira Xisto, Lisbon.

Baart, Jan. 1988. Faiança portuguesa, 1600–1660. Um estudo sobre achados e colecções de museus. In *Portugueses em Amesterdão. 1600–1680*, edited by Renée Kistemaker and Tirtsah Levie, pp. 18–24. Amsterdams Historisch Museum, De Bataafsche Leeuw, Amsterdam.

Bebiano, Rui. 1987. D. *João V—Poder e espectáculo*. Estante, Aveiro.

Bernardes, Padre Manuel. 1699. Armas da castidade. Oficina de Miguel Deslandes, Lisbon.

Biedermann, Zoltán. 2014. Imperial Reflections: China, Rome and the Spatial Logics of History in the Ásia of João de Barros. In *Empires en marche*, edited by Dejanirah Couto and François Lachaud. École Française d'Extrême-Orient, Paris.

Bluteau, Rafael. 1712–1728. *Vocabulario portuguez e latino.* 10 vols. Colégio das Artes da Companhia de Jesus, Coimbra; Oficina de Pascoal da Silva, Lisbon.

Casimiro, Tânia Manuel. 2011. *Portuguese Faience in England and Ireland.* BAR International Series 2301, Oxford.

Coelho, Rui Gomes. 2009. Comportamentos de resistência à integração colonial na Amazónia portuguesa (Século XVIII). *Anais de História de Além-Mar* 10:129–184.

———. 2010. A intervenção arqueológica no largo António Joaquim Correia. Contributo para o estudo da vida quotidiana em Setúbal no tempo do Grande Terramoto. [Archaeological Intervention at the António Joaquim Correia Square. Contribution to the Study of Daily Life in Setúbal at the Time of the Great Earthquake]. MA thesis, History Department, New University of Lisbon, Lisbon.

Corbin, Alain. 1982. *Le miasme et la jonquille.* Flammarion.

Cox, Thomas, and Cox Macro 2007 [1701]. *Relação do Reino de Portugal,* coordinated by Maria Leonor Machado de Sousa. National Library, Lisbon.

Domingues, Ângela. 2000. *Quando os índios eram vassalos. Colonização e relações de poder no Norte do Brasil na segunda metade dos Século XVIII.* Comissão Nacional para as Comemorações dos Descobrimentos Portugueses, Lisbon.

Duhart, Frédéric. 2001. *Habiter et consommer à Bayonne au XVIIIe siècle.* L'Harmattan.

Faria, Manuel Severim de. 1791 [1625]. *Notícias de Portugal,* Tomo I., Oficina de António Gomes, Lisbon.

Fernandes, Isabel Maria. 2013. A loiça preta em Portugal: Estudo histórico, modos de Fazer e usar (Black Pottery in Portugal: Historical Research, Forms of Production and Use). Unpublished PhD dissertation, History Department, Minho University, Braga.

Ferreira, Luís Gomes. 1735. *Erário Mineral.* Oficina de Miguel Rodrigues, Lisbon.

Foucault, Michel. 2000. *Power: Essential Works of Foucault,* Vol. 3 (1954–1984). New Press, New York.

Freire, Simão Antunes. 1731. Soneto. In *Rimas sonoras. Segunda parte das obras académicas de Simeam Antunes Freire,* pp. 7–37. Lisbon, Oficina Augustiniana, Lisbon.

Gell, Alfred. 1992. The Technology of Enchantment and the Enchantment of Technology. In *Anthropology, Art, and Aesthetics,* edited by Jeremy Coote and Anthony Shelton, pp. 40–63. Oxford University Press, Oxford.

———. 1998. *Art and Agency. An Anthropological Theory of Art.* Oxford University Press, Oxford and New York.

Graeber, David. 2004. *Fragments of an Anarchist Anthropology.* Prickly Paradigm Press, Chicago.

Hatherly, Ana. 1993. As misteriosas portas da ilusão: A propósito do imaginário piedoso em Sóror Maria do Céu e Josefa de Óbidos. In *Josefa de Óbidos e o tempo barroco,* organized by Vítor Serrão, pp. 71–85. TLP-Instituto Português do Património Cultural, Lisbon.

Hespanha, António Manuel. 1994. *As vésperas do Leviathan. Instituições e poder político. Portugal—Séc. XVII.* Almedina, Coimbra.

———. 2006. As cores e a instituição da ordem no mundo de antigo regime. *Philosophica* 27:69–86.

Henriques, Francisco da Fonseca. 1726. *Aquilegio medicinal. Em que se dá noticia das agoas de caldas, de fontes, rios, poços, lagoas, e cisternas, do Reyno de Portugal, e dos Algarves, que ou pelas virtudes medicinaes, que tem, ou por outra alguma singularidade, são dignas de particular memoria.* Oficina da Música, Lisbon.

———. 1731. *Ancora medicinal, para conservar a vida com saude*. Oficina Augustiniana, Lisbon.

Jesus, D. Fr. José de Santa Maria de Jesus. 1735. *Brados do pastor às suas ovelhas. Obra espiritual dividida em duas partes*. Oficina de Manuel Fernandes da Costa, Lisbon.

Lavanha, João Baptista. 1622. *Viagem da Católica real magestade del rey D. Filipe II N. S. ao Reino de Portugal e relação do solene recebimento que nele se lhe fez*. Tomás Junti, Madrid.

Lapa, João Inácio Ferreira. 1868. Relatório sobre os processos da vinificação nos principaes centros vinhateiros do sul do Reino. *O Archivo Rural* 11:265–272.

Luís, Friar Manuel de São. 1731. *Instrucções moraes, e ascéticas deduzidas da vida, morte e virtudes da venerável Madre Francisca do Livramento abbadessa que foy no mosteiro de N. S. da Esperança da cidade de Ponta-delgada, ilha de S. Miguel*. Livro II. Oficina Augustiniana, Lisbon.

Machado, José Timóteo Montalvão. 1965. No centenário do I recenceamento populacional português. *Revista do Centro de Estudos Demográficos* 16:81–107.

Monteiro, João Pedro. 1994. A influência oriental na cerâmica portuguesa do séc. XVII. In *A Influência oriental na cerâmica portuguesa do séc. XVII*, edited by Maria Antónia Pinto de Matos and João Pedro Monteiro, pp. 18–55. Museu Nacional do Azulejo, Lisbon.

Morais, Pedro José Supico de. 1761. *Collecção moral de apothegmas, ou ditos agudos, e sentenciosos, novamente impressa, correcta e illustrada*. Parte II. Oficina de Francisco de Oliveira, Coimbra.

Moreira, Rafael. 1998. Cultura material e visual. In *História da expansão portuguesa*, directed by Francisco Bethencourt and Kirti Chaudhuri (dirs.), Vol. 1, pp. 455–487. Temas e Debates, Lisbon.

Newitt, Malyn. 2005. *A History of the Portuguese Overseas Expansion, 1400–1668*. Routledge, London and New York.

Pastoureau, Michel. 2001. *Blue. The History of a Color*. Princeton University Press, Princeton and Oxford.

Sebastian, Luís. 2011. A produção oleira de faiança em Portugal (Séculos XVI–XVIII) [Faience Production in Portugal (16th–18th Centuries)]. Unpublished PhD dissertation, History Department, New University of Lisbon, Lisbon.

Semedo, João Curvo. 1697. *Polyanthea medicinal. Notícias galenicas, e chymicas, repartidas em tres tratados*. Oficina de Miguel Deslandes, Lisbon.

Senos, Nuno. 2002. *O Paço da Ribeira: 1501–1581*. Notícias Editorial, Lisbon.

Soares, Joaquina. 2000. Arqueologia urbana em Setúbal: Problemas e contribuições. In *Actas do encontro sobre arqueologia da Arrábida*, pp. 101–130. Instituto Português de Arqueologia, Lisbon.

Simões, João Miguel dos Santos. 1997. *Azulejaria em Portugal no século XVII*. Vol. 1. *Tipologia*. Lisbon, Calouste Gulbenkian Foundation.

Sousa, Friar Luís de. 1921 [1619]. *Vida de D. Frei Bartolomeu dos Mártires*. Livrarias Aillaud e Bertrand, Paris and Lisbon.

Souza, Laura de Mello e. 2003. *The Devil and the Land of the Holy Cross. Witchcraft, Slavery, and Popular Religion in Colonial Brazil*. University of Texas Press, Austin.

Tavares, Rui. 2005. *O pequeno livro do Grande Terramoto*. Tinta-da-China, Lisbon.

Vasconcellos, Carolina Michaëlis de. 1957 [1921]. *Algumas palavras a respeito de púcaros de Portugal*. Revista Ocidente, Lisbon.

Velho, Father Manuel. 1730. *Cartas directivas e doutrinaes. Respostas de hua religiosa capucha, e reformada, a outra Freira, que mostrava querer reformarse.* Oficina de António Pedroso Galrão, Lisbon.

Vieira, António. 1744 [1652]. *Arte de furtar, espelho de enganos, theatro de verdades, mostrador de horas minguadas, gazua geral dos Reynos de Portugal.* Oficina de Martinho Schagen, Amsterdam.

CHAPTER SIX

Quotidian Agency and Imperial Agendas

A Study of Andean Middle Horizon Huamanga Ceramics

Brittany Fullen

In the endeavor to construct ceramic styles, researchers identify formal attributes representing commonalities to draw distinctions that create different stylistic groupings. Through this process, we whittle down the variation we observe in our assemblages to get at analytically manageable and discussable configurations. In this chapter, I propose that we invert the usual process by placing our questions about how ceramics were used in the past as a starting point, not an ending one. When we analyze ceramics based on formal properties, such as shape, size, slip, and decoration, we arrive at categories that may have been unrecognizable to the people who actually used them. What if flexibility in decoration was an important quality in a ceramic style? Although this may seem counterintuitive, if ceramics were used as a physical means of negotiating identity, for example, then flexibility in decoration may be a key to understanding the style and the social relationships embedded within them.

Ceramic taxonomy has been integral to Andean studies past and present. Through the identification of formal properties, it has constituted the baseline for spatial and temporal designations of cultural units and identities. Using rigid guidelines, researchers have organized diverse material assemblages to observe continuity, transformation, collapse, and interaction through time and space. However, these static classifications are limited tools for understanding active and dynamic social relations. To further complicate the situation, many Andean style/type categories were created with whole ceramics from mortuary contexts with relatively little known provenience; when vessels were from provenienced locations, they tended to comprise relatively small samples.

Ceramics are not just passive indicators—receiving but not acting back—of events, people, and ideas in which they are final products of the outcomes of human interactions (e.g., Gell 1992, 1998; Ingold 2007). Rather, ceramics have agential, and even animistic, qualities that shape the social interactions and behavior of humans (e.g., Alberti and Marshall 2009). Viveiros de Castro's perspectivism (1998, 2004) can be extended beyond humans and animals to objects, which may be seen to have their own kinds of agency or intentions. Gosden states that "styles of objects set up universes of their own into which people need to fit" (Gosden 2005:194). Objects channel human intentions; they have life cycles, and they have the ability to facilitate the creation of more objects within their style. Because objects are endowed with power and agency, we are socialized into a material world from the moment of our births.

An agency approach to ceramic style may be a more appropriate thinking tool than a static, typological grouping. Agency-based studies have focused on special, exotic, and ritual objects (e.g., Can, this volume; Gell 1992; Gruner this volume; Mills and Ferguson 2008; Weismantel 2013; Zedeño 2008), as well as the mundane, everyday things that comprise the majority of human-material interaction (e.g., Alberti 2012; Chiykowski this volume; Coelho this volume; Gell 1998; Gosden 2005; Groleau 2009; Viveiros de Castro 1998, 2004; Young-Wolfe this volume). Although Andean scholars have extensively evaluated exotic objects to uncover information about past lifeways and interactions (e.g., Herring 2010; Menzel 1964, 1968), it is equally if not more important to focus on artifacts that interacted with people on a regular basis—the objects that mediated, shaped, and reinforced past people's identities, social relationships, and cultural worlds.

In this chapter, I focus on the Middle Horizon (AD 600–1000) Huamanga (Wamanga) pottery of pre-Hispanic Peru to illustrate the beneficial application of agency theory to ceramic style studies and to investigate how the spread of a ceramic style can be related to the expansion of an empire. My observations are focused on a collection of ceramics excavated between 1999 and 2003 at the Wari site of Conchopata, located in the Ayacucho Basin in the Wari heartland. I conduct a visual analysis based on photographs, and I compare the Conchopata assemblage with ceramics from other sites in presumed provincial territories in order to explore arguments about variation and regionalization.

Object Agency: A Gellian Perspective

Alfred Gell (1998) offers an insightful framework from which to understand the role objects play in human interaction. For Gell, agency revolves around the object found within the nexus of social interaction. When the object mediates relations among social others, the object, or (as he calls it) *index*, allows the observer to *abduct* (make inferences) "the intentions or capabilities of another person" (Gell 1998:13). Two things are important here: 1) objects are not passive materials but active agents within the social world; and 2) object agency differs from human agency. Gell argues that beings with intentionality, namely humans, have *primary agency*, whereas things lacking intentionality, yet still having the ability to bring about causal action, have *secondary agency*. It is important to remember that secondary agency is "relational and context-dependent" (Gell 1998:22). Agency is not expressed until action takes place, which requires the interaction of an *agent* and a *patient*. An agent brings about action, while a patient is the one causally affected. Patients are also potential agents. However, they do not necessarily take a passive role but may resist the agency of the agent.

Agents and patients can be any combination of artists, indexes, recipients, and prototypes. An artist creates an index.[1] An index is the material entity through which the agency of either the artist and/or the recipient is abducted. Recipients are the intended audience for the index and are those on whom the index exerts its agency. However, as mentioned, recipients may use the index to exert agency on others in the forms of patrons or an audience. Prototypes are the ideal blueprint for the index and exert their agency on the artist when the artist is creating an object. It should be noted that though there may be some conceptualized and standardized ideal form, artists are just as much (or arguably more) affected by the process of creating objects (Chiykowski, this volume) and by the use and interaction with the indexes. As Ingold (2007) argues, it is through working with or through materials that we (and our actions) develop, no matter how seemingly prototypical. Artists are affected by previous works they have either encountered or created (Gosden 2005).

Objects can be instruments of social agency, as well as the outcome of it. The concept of objects as the outcome of human action is readily accepted in the field of archaeology, so I will only discuss objects acting as instruments of social agency. All artifacts exist as the product of human agency, and as such, have the ability to index their maker through what Gell calls the *abduction of agency*. However, this is not always referenced by objects because the link between maker and object can be forgotten, concealed, or may have never been known in the first place. For example, when objects were imported from distant regions, their place or group of origin may have been recognized but not their individual maker. So alternatively, objects can rely upon a second abduction of agency. As no objects are made for no reason, the object may also pull its agency from its intended recipient(s). Because objects are embedded within social relations from the moment they are created, they have the ability to allow us as archaeologists to tap into social relationships. This will be explored more thoroughly later in the chapter by examining the feasting context of the Wari.

Although all objects are created by the artist's agency, that does not mean that these indexes cannot exert agency onto the artist. Objects often causally affect artists in the form of traditional knowledge (Gell 1998:29). This can include production processes and style. As Gosden argues, "objects produced within a recognizable set of forms and styles have influences on the ways in which people make and use them" (Gosden 2005:194). In other words, you do not have style without traditional knowledge. The artist must take into account pre-existing notions of a style (such as a Huamanga pot) because the style has a physical manifestation. Artists must have some fidelity to the style if they want people to recognize and understand their work. By fidelity I do not mean exact replication, as we have a plethora of examples displaying the variability expressed by potters whose agency has been incorporated into the process. However, the new vessel must contain enough relevant attributes in common so that references would be upheld. Therefore, "regular and generally small modifications" (Gosden 2005:195) permit artists to negotiate through a physical medium. By focusing on the active nature of the ceramics, the researcher can get an idea of the constraints on the process coming from the style, the physical limitations of the ceramic, and the artist's own intentions (for a good example see Chiykowski this volume). Additionally, artists do not just create one "work," but a series of them throughout their careers. In the construction of each object, they are influenced not only by that process in action but by all objects they have ever created. Gell specifies that these indexes then go on to exert agency onto the recipient(s) in which the individual or group submits to the object's "power, appeal, or fascination . . . [and is]

responding to the agency inherent in the index. This agency may be physical, spiritual, political . . . as well as 'aesthetic'" (Gell 1998:31).

There are nearly an infinite number of possible social interactions, but in all of them, the object is the pivot of the nexus; intentional agency and patient-hood fall just outside. We can think of objects as proxies that act as extensions of the patron and/or the artist. At the same time, Gell argues that the index can also be a bundle incorporating the patient-recipient, which can then be grasped and manipulated by others (Gell 1998:37). So how does this translate to our topic of ceramic agency and the construction of ceramic styles? Thus far we have been talking in terms of this simple formula: Index-Agent → Recipient-Patient. This could describe many different interactions a person could have with a Huamanga ceramic vessel. But the nexus we are discussing really looks like: ([(Prototype-Agent) → Artist-Agent] → Index-Agent) → Patient-Recipient. In the latter formula, the origin of the abducted agency is expressed in the index agent. The index is not acting on the recipient alone but is the carrier of the causal agency and mediates agency effect on the patient-recipient. This is done by the recognition on the part of the patient that the object is a "made thing" and thus is the outcome of an artist's intentions. The object is a patient in relation to the artist and an agent with respect to the recipient. The artist also falls into this scenario, as she/he is affected by the agency of the prototype and on some level is subordinate to the prototype.

In the Middle Horizon Andean example here, the prototype is the Huamanga style, and the artist is a potter.[2] The potter on some level is subordinate to the ceramics' style assemblage in existence. If the intention were to incorporate the same ideas, identities, and so on within the ceramics, the potter would have to maintain fidelity to a pre-existing prototype. The index would be the actual ceramic produced, and the recipient would be the individual or group who would use the ceramic in daily life. For the Wari, the highly standardized corporate style would be utilized by administrators of the empire to introduce and indoctrinate new members through the controlled message the ceramics contained. Those receiving the index would be aware of the agency behind the production of the object and in turn would be making a statement through the choice to use them in their daily activities. They could be signaling their inclusion or their desired inclusion through the use of the ceramics. However, people could have also chosen to use ceramics with subverted messages as a form of resistance to the empire. Through the use of these concepts, we can demonstrate the way in which object agency can infiltrate and influence daily life.

Middle Horizon Wari Entanglements

Because ceramic styles had meaning to the people who created and interacted with them, during every interaction with the object, the individual or group would be reminded, consciously or unconsciously, of the meaning, message, or idea reinforcing their conceptualization of their social world and their place within it. This is integral to the archaeological process of ceramic style construction. In ceramic style studies, the question of "how ceramics operated in past societies" usually becomes a concern only after the style has been defined. Generally, the idea is that in order to understand what a ceramic *did*, we first must decide what the ceramic *was*. By contrast, I suggest reversing the order of the questions; to understand what a ceramic *was* we first need to understand what a ceramic *did*.

I adopt this approach in the investigation of Middle Horizon Wari Huamanga ceramics. The Wari culture and people of prehistoric Peru had their capital in the south-central highlands, in the Department of Ayacucho. Their empire spread west to the coast, south toward the border of modern-day Bolivia, north up through the Cajamarca region, and included much of the coast and highlands. The quotidian Huamanga style has been identified as a Wari ceramic type that is diagnostic to the Ayacucho heartland around the capital city of Huari.[3] Researchers have identified Huamanga ceramics outside the Ayacucho region along with Huamanga-like local varieties. There is a great degree of variation found within Huamanga styles within the heartland (Anders 1986, 1989) and in the provincial regions (Owen 2007, 2010). By examining ceramic objects as secondary agents, I am attempting to refocus the discussions surrounding Huamanga and Huamanga-like styles.

An underlying assumption to this discussion revolves around the political organization of the Wari. My argument is grounded in the idea that Wari was an empire with Huari as its capital and the Ayacucho region as its heartland (Glowacki and Malpass 2003; Isbell 2001; Schreiber 1992, 2001, 2012). As the Wari expanded their realm of influence across the Andes, they created a mosaic of direct and indirect control, skipping some valleys altogether. Wari expansion entailed some form of a governing body with an agenda and clear concept of Wari identity. Although a religious and ideological doctrine likely was perpetuated by the state, it may have been in Wari's best interests to attempt to incorporate local and regional ethnic groups through appropriation of local core beliefs, concepts, and/or identity. I am reworking (to an extent) the ceramic description of Huamanga based on a large ceramic collection from Conchopata. To do so, I am concerned with how and why people of the Wari Empire would have used this quotidian ceramic.[4]

In the Andes, notions of animism (e.g., Groleau 2009; Viveiros de Castro 1998, 2004) and active agency of inanimate objects is prevalent. From sacred places on the landscape, such as mountain spirit *apus* (ancestors), *pacarinas* (ancestor origin), and *upaimarcas* (resting places), to active, "living" portable goods, such as ceramic vessels, the Andes was an environment in which things and people had an intimate and mutually affective relationship (Glowacki and Malpass 2003; McEwan and Williams 2012; Williams and Nash 2006). In the Middle Horizon, we find evidence of *camay* (life force) in the smashed remains of the ceramic offering tradition. The ceramic offering tradition can involve plain, as well as fancy ceramics, but it is identified by the involvement of Conchopata faceneck jars. These ceramics are characterized by faces molded into the neck and sometimes shoulder of large—and in this case—oversized jars. The oversize urns and faceneck jars contain a complex state iconography (Nash 2012) of figures and geometric patterns and were used in feasting contexts. Faceneck jars may have been stand-in representations of hosts, dignitaries, ancestors, or individuals who could not attend the feast. If these vessels were participants in the feast and not just serving vessels, then the wealth, food, and drinks physically emanating from their body directly provided the sustenance for the other guests. It was no longer a display and inference of support but an embodied act and reminder of what a patron provided to his/her clients.

Later, many of these vessels were destroyed by multiple direct blows to the images or anthropomorphic lugs at the termination of the gathering. They were then interred along with other vessels into pits, left on the floor where they were broken, or deposited in other rooms. By breaking the pot, it released the life force inside the ceramic to reanimate elsewhere in the world. This activity has been associated with commemoration of the dead and abandonment practices that terminate the use of the room(s) or structure(s). Although I do not at this point consider Huamanga ceramics to be animated, an object's ability to act, influence, and have power was a core concept across the Andean realm.

The iconography depicted on these vessels comprises only some of the ceramic style variation observed in Wari contexts. Quotidian Huamanga bowls and cups have been recovered from these feasting contexts. Nash has argued that "the abstract icons that appear on tableware—cups and bowls—may have functioned in part as heraldry and identified family, occupation, rank, title, or patron" (Nash 2012:89). It seems that if the main serving ware had state-controlled corporate iconography, it is possible that the less well-made and less-performatively visible vessels could have displayed local conceptual inclusion to the feast. Just as the social milieu at feasts incorporated at times varying social classes of individuals, with different agendas and roles to play, so, too, could the ceramics be a mixture of ideas and references (especially if they were subtle). It would generate a more inclusive atmosphere at the gathering and would not interfere (and possibly enhance) the goal of the assembly.

Feasting was an activity that occurred on a regular basis in the Wari Empire; it was hosted by elites who intended to bring people together to serve the needs of several fundamental institutions. Feasting, in many ways, was the "social glue" of the Wari Empire. Nash (2012:82) notes that feasting "can be self-perpetuating mechanisms for maintaining the relationships they establish; in many societies they create networks of mutual obligations between hosts and guests with far-reaching implications. In other words, feasting can be a significant institution that binds people together, defines their relationships, and drives economic production."

As elites routinely practiced feasts, a feasting infrastructure must have been in place that required organization and carefully scheduled time to pull together larger events during the Middle Horizon (Nash 2012:86). Elites would have had dedicated spaces to hold feasts and would have had institutionalized contributions from "followers" or hired staff to collect, prepare, and carry out the duties necessary for conducting a feast start to finish. If, however, it was not possible to have a reliable labor force dedicated solely to feasting, a network of reciprocal obligations could accomplish the same end result. In either form, feasts create situations in which relationships can be negotiated and fomented among many different members of society (and even outside the group). Even today, obligations are a fundamental underpinning to daily society, as favors and other gestures are never free.

The ritual smashing of vessels also occurs in state administrative practices during the Middle Horizon. Feasts provided an opportunity for political integration and negotiation of power, prestige, status, and identity between Wari administrators, local populations, and neighboring allies ([or potential allies] Cook and Glowacki 2003). They can be used to "promote state agendas, demonstrate state success through pageantry, and reinforce state ideologies through ceremonies or performances. From a political perspective feasts are opportunities for people to assert power over others by creating obligations, winning the admiration of followers, or outcompeting rivals" (Nash 2012:84).

Feasts utilized several different drinking and serving vessels, including oversized serving vessels (urns, faceneck jars, and tumblers), as well as more individually sized lyre-shaped cups, tumblers and banded tumblers, open straight-sided bowls, regular straight-sided cups, and regular incurving bowls. Many of these are present within the Conchopata collection discussed below. Cook and Glowacki's (2003) feasting studies conducted at Moraduchayuq and Azángaro in the heartland and Pikillacta and Huaro in a provincial region found that continuity in vessel styles linked the vessel forms and the imagery to the ritual and political administrative activities of the state. Commensal displays of generosity, power, status, and political prowess implies an imperial Wari identity that was arguably reinforced through the pomp and circumstance of the administrative ritual and political process of feasting and hospitality.

Dorothy Menzel (1964, 1968) defined most Wari ceramic styles in her seminal works. Initially, she created ceramic styles based on a relatively small sample size.[5] However, after more than forty years of additional excavation and study, and the ever-increasing number of radiocarbon dates, it is clear that these styles need to be adjusted and reworked (Isbell and Knobloch 2006, 2009). Additionally, there are some ceramic styles that are commonly referenced by Andean scholars today that lack an extensive published description. This becomes problematic because scholars believe they are talking to each other about the same thing, when in actuality they could be operating with categories and styles that are quite different. Some discrepancies among styles and types may be due to differences in the conceptual approach to the process. In descriptions of the Huamanga style, some authors may be taking an iconographic approach to the understanding of these pots (e.g., Owen 2007). However, in using and describing the ceramics as a quotidian ware, other scholars are taking an approach that is focused on the way in which the vessels were used (e.g., Anders 1989). These two different approaches come with different sets of preconceived notions as to what variation could be included or tolerated and what should be excluded. Furthermore, some researchers have used Huamanga quotidian ware as a residual category for items that do not fit into other categories or were not particularly noteworthy, exotic, or unusual.

The Huamanga Ceramic Style

Huamanga ceramics were initially identified and separated into their own category by Luis Lumbreras (1974) and Mario Benavides (1965). Using a type-variety approach, Lumbreras and Benavides laid out a very specific and lengthy description of Huamanga and its motifs. Menzel had not separated the style out in her earlier style studies in 1964 and 1968 but instead lumped it in with other styles.[6] Since then, archaeologists have been writing about Huamanga ceramics but have never reconciled the differences between the stylistic definition and the type-variety approach (William Isbell, personal communication 2012).

Huamanga is a quotidian ceramic style considered to originate in the Ayacucho region, the heartland of the Wari. Some scholars have documented Huamanga's appearance in other regions up to 370 kilometers from the Ayacucho Valley (Glowacki 1996; Owen 2010). As scholars continue to grapple with what the Huamanga ceramic style is, how it was used, and how people related to it, discussions have shifted to ideas of regionalization (Anders 1989; Owen 2010). Anders argued for regionalization within the Ayacucho Basin and cited the maintenance of pre-existing local ethnic identities from the Early Intermediate Period (AD 370–600) by using Huamanga ceramics as a tool of resistance to Huari homogenization. Owen (2007) argues that observed variation is the result of being a less institutionalized ceramic style produced on the household or village level. Part of this inference seems to derive from the idea that Huamanga vessels are "less iconographically loaded" (Owen 2010:61).

Often, the Huamanga style has been described as a lower-quality version of finer ware traditions or poorly made ceramics in general. Most are decorated but considered to be "less iconographically complex than vessels on the formal end of the spectrum" (Owen 2007:289). From descriptions provided by Isbell (1977) and Anders (1986), the Huamanga ceramics generally exhibit color variation on the surface of the vessel and in the paste. Vessel surface colors include buff, dull orange, and medium reddish orange (Anders 1986; Isbell 1977), as well as a red color (Owen 2007). The ceramics are typically sand tempered with visible small inclusions that sometimes include mica, quartz, and feldspar. An open bowl (Anders 1986; Isbell 1977; Owen 2010) and a somewhat restricted jar form are common vessel shapes, and the texture of the ware varies. The vessel surfaces often have a dull matte finish (Anders 1989; Isbell 1977) exhibiting specific colors (buff and red slipped with black, red, white/cream, orange, gray, and dark red/reddish brown paint) and color combinations (Isbell 1977; Owen 2007). Decoration consists of painted motifs usually geometric in nature with a strong preference toward symmetry.

Almost all painted vessels include black paint and minimally a line (Owen 2007) and were organized into bounded space and modalities. Decorations often incorporated repetitive simple patterns that sometimes

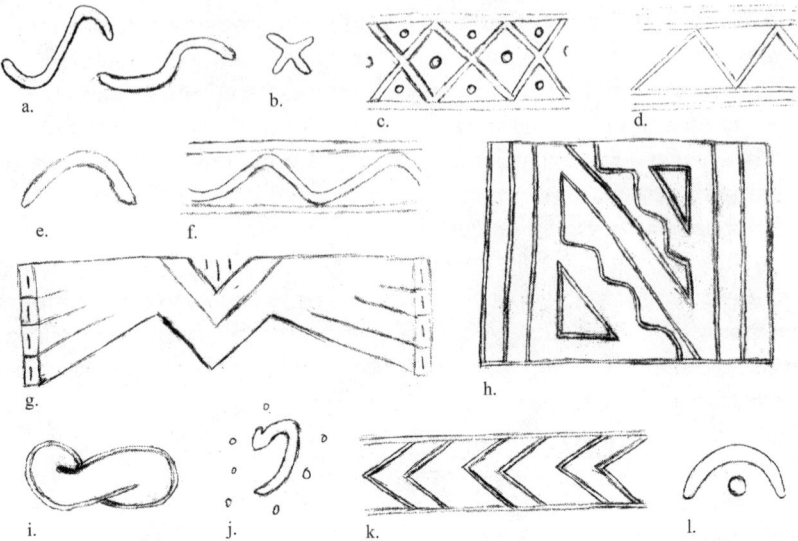

Figure 6.1. Huamanga motifs: A) lazy S; B) lazy X; C) lozenge band; D) zigzag band; E) shallow arc; F) wavy line; G) feathered wing; H) escalonados; I) double recurved ray; J) hook and dots; K) chevron; L) semi-circle with dot. Drawings by author and drafted by Anne Hull.

included filled color segments. The most common design is a band located along the top of the vessel at the rim, often marked by one or more lines running along the bottom of the band. Geometric figures, such as lazy S and X, lozenge band, and zigzag bands are often located in these continuous bands (Figure 6.1). These are found on the exterior of bowls except for the straight-sided flaring bowls, also called *escudillas* ([Figure 6.2] Owen 2007). The next most common design pattern employs continuous bands divided into registers that alternate or repeat (containing one or more designs followed by empty space). These bands can also exhibit a checkerboard motif. Additionally, the bowls can be divided vertically by pendants/bars that run from rim to base. Shallow arcs, wavy lines, *escalonados* (a stepped stair-like geometric design), and "feathered wings" also appear on many vessels (see Figures 6.1 and 6.2). Some of these motifs were often set into mirrored pairs. Owen argues that

> Huamanga ceramics are essentially the more rustic end of the range of variation of Wari ceramics, losing many of the diagnostic features of the formal styles such as Chakipampa and Viñaque and incorporating

additional, less systematically controlled motifs and concepts that are not found in the formal rules. This makes it difficult to link most Huamanga vessels or assemblages to any one of the formal Wari styles. At the same time, Huamanga ceramics are not sharply distinguishable from the lower-quality variants of Chakipampa, Ocros, Viñaque, and other formal styles, and separating them involves an arbitrary division of a continuous range of variation. (Owen 2007:305)

Although Huamanga ceramics do not lend themselves to a tight formal definition, researchers can and do recognize the concept of Huamanga in the archaeological assemblage. This brings me to the collection from Conchopata.

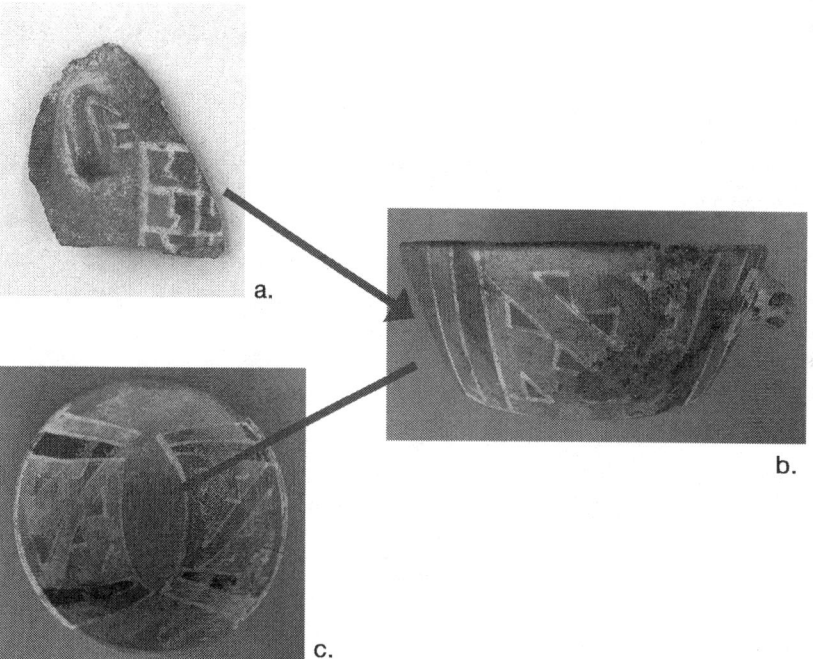

Figure 6.2. A) Molded face with escalonados cheek design (Courtesy of the Division of Anthropology; YPM [ANT. 212201], Peabody Museum of Natural History, Yale University, New Haven, Connecticut). Photograph by Patricia J. Knobloch, William H. Isbell, Brittany A. Fullen, and Edward P. Zegarra; B) Escudilla with zoomorphic lug and escalonados motif. Photograph by William H. Isbell; C) Escudilla with escalonados motif. Photograph by William H. Isbell.

Case Study: Middle Horizon Wari at Conchopata

The ceramic collection used in this study is from the site of Conchopata located in the Ayacucho region about ten kilometers from the capital city, Huari. It is considered to be an important site of ceramic production and politico-religious activities in the Ayacucho Valley; it is second only to the urban capital site of Huari itself (Cook and Glowacki 2003). William H. Isbell and members of the Conchopata Archaeological Project excavated the collection between 1999 and 2003. Several tons of ceramics were unearthed, photographed, and cataloged. The data examined here comes from high-quality digital photographs of individual sherds, as well as semi-complete and complete vessels. The photographs are organized into 2,813 groups of individual and multiple specimens. I examined vessel form, iconographic motifs, and decorative spatial arrangement. I compared the collection to specimens evidenced by published photographs and drawings from neighboring regions (Anders 1986, 1989; Bauer 2002; Glowacki 1996; Owen 2007, 2010).

There is a variety of styles within the collection (in addition to Huamanga): Chakipampa, Ocros, Conchopata, Robles Moqo, Viñaque, Viñaque-associated, and Black-Incised (as defined by Menzel 1964). Most of the ceramics photographed were sherds, many of which were only five to eight centimeters across. Additionally, a large proportion of the collection consisted of undiagnostic sherds that appeared to be unpainted and unslipped (at least with a different color slip than the natural buff, clay color). This could reflect the fact that: 1) most vessels were not heavily decorated from rim to base, but around the neck and shoulder on jars and the rim on bowls; or 2) some of the sherds represent ceramic firing production failures (cracked or exploded vessels).

Most Huamanga ceramics are open, straight-sided flaring bowls with decoration found on only one side (mainly the interior). There are three typical design styles: 1) rim bands featuring geometric patterns; 2) vertical pendants; and 3) mirrored patterns that cover approximately 25 percent of the interior (50 percent for the pair) and are located opposite each other with empty space between them. The designs are almost always banded or bounded in some way and have thick black outlines. On some vessels, potters combined these principles, including figural representations, in addition to or instead of geometric motifs. Jars include faceneck fragments and molds for making the face. Other non-Huamanga, globular jars are undecorated or highly burnished, black-slipped, faceneck jars.

Some lyre cups, some *keros* (tall drinking cups), one canteen, and some oversized urns in the collection do not represent Huamanga style.

Many miniatures were recovered whole. Although most miniatures are undecorated, a few are decorated in the Chakipampa, Ocros, or Huamanga styles.

Most vessels in the collection are plain slipped (buff to orange) or red slipped, with a few highly polished black-slipped vessels. The plain/buff vessels and the red-slipped vessels are mainly decorated with black paint, which is followed in popularity by cream/white. Red is the only other color commonly used in the Huamanga style, while vibrant purple, orange, and gray are used on several of the Chakipampa and Ocros vessels. Motifs for Huamanga consist of the following: black lines and waves, single and multiple bands of solid colors, bands with lazy S and double recurved rays (like infinity signs), feathered wings, escalonados, dots and hooks, and chevrons (see Figures 6.1 and 6.2). Although Owen (2007) cited comb motifs as common on Ayacucho-area ceramic vessels, I did not find this motif in the Conchopata assemblage.

I observed an interesting correlation between motifs and their placement on the faceneck jars, open bowls with anthropomorphic and zoomorphic lugs, and the open bowls with geometric patterns. Most of the faceneck jars with anthropomorphic modeling exhibit a geometric design on the cheeks and a chevron or banded pattern on the hat/headband of the individual or above the head around the rim of the jar. Additionally, all anthropomorphic and zoomorphic figures had either a decorative geometric collar/pectoral or a black lined or banded outline under the chin. Other figural and geometric forms were located on the shoulder, and just below the shoulder, that apparently represent features of dress—almost certainly a tunic.

On open bowls, anthropomorphic and zoomorphic lugs often are located on or near the rim on the outside of the vessel. When anthropomorphic lugs are present, there often are mixtures of bands and the geometric collars/pectorals surrounding the head. In addition, a band or chevron band circles the rim. Bowls with zoomorphic lugs, by contrast, exhibit a white background with black or red dots, and at times semicircles, located on the face, neck, and (when present) body of the creature (see Figure 6.2). This is surrounded by a black outline. There are also bands located around the rim.

On open, straight-sided, flaring bowls, chevron bands, and other geometric bands run around the interior of the rim (occasionally on the outside only). On some vessels, bands extend vertically, running from the rim to the base on the interior in a mirrored manner. Additionally, the stepped-geometric block appears in a set, and each covers about 25 percent of the interior of the vessel where they are placed across from each other. At times the stepped motif is bounded by plain vertical lines and bands on either side (Figure 6.2).

Clearly there is a recognizable order and compartmentalization to the ceramic motifs as seen in the outlining and bounding of space and designs. An element of control is exhibited at all times, and potters often mixed in whimsical and supernatural constructions (Cook 2012; Nash 2012). Things have their places on the ceramics, so what we might be seeing on these nonfigural plainware is an abstraction of the fancy ware. Although there are a large number of possible combinations of design and color, potters selected the same ones across the different vessel forms (bowls and jars).

On the bowls previously discussed, potters employed a deconstructed, geometric abstract reference to human decoration in the same locales as those found on anthropomorphic and zoomorphic vessels. By occupying the same space, and by using identical or similar patterns and color combinations, potters linked concepts found within the fancy and ceremonial styles to those of the everyday. When comparing this to the images Owen (2007, 2010) provided for Beringa, Glowacki (1996) for Pikillacta, Bauer (2002) for Cuzco, and Anders (1986, 1989) for Azángaro, this trend seems to hold based on the few photos of vessels and fragments published. However, the motifs potters chose varied considerably. The escalonados is only found in the Azángaro and Conchopata sites, both located in the Department of Ayacucho. The Araway/Arahuay style of Cuzco and the Pikillacta styles both emphasize dots, wavy messy lines in bands, and hooks; both of these sites are in the Cuzco region. The Conchopata assemblage does have some hook-like lines, but they are semicircle or crescent shaped and often have a dot in the center of the curve. Conchopata sherds also commonly had black dots, but they were usually associated with a zoomorphic lug or figural design. Beringa and Conchopata assemblages contain feathered wing designs, but for the majority of images published by Owen, these designs consist of a diamond lattice-like band (lozenge style) and lazy S/double recurved rays in bands. Although the double recurved rays were common in bands at Conchopata, they were usually located on the exterior of incurving bowls along the rim. By contrast, the Beringa bands were located on the interior of escudilla bowls. It should be noted that comparison with published materials was limited because there were not many photos of faceneck jars or bowls with lugs to compare with the Conchopata collection. Because publication limited the selection of specimens, they might not have exhibited all the trends in the extensive Conchopata collection. However, it should be recognized that there are motif groupings regionally organized (although at times overlapping). This could relate to Isbell and Knobloch's (2006, 2009) arguments about ethnic identity.

Discussion

Travel in the Andes was difficult, and it seems unlikely that most Wari subjects could journey to the heart of the Wari Empire. Furthermore, not all areas had Wari colonists or administrators. Nonetheless, people seem to have been using Wari objects to mediate social relationships and identities. As Gell has shown, objects have the ability to move between people while expressing ideas, sentiments, and agential purpose without the in-person involvement of all human participants at all moments. Because the Huamanga style is associated with Wari, Huamanga ceramics could have extended the agency of the Wari across space, serving as ambassadors for the empire. Ceramics are a highly visible, portable, and malleable medium with which people can engage. The style spread rapidly in an environment that consisted of a mosaic of valleys under direct and indirect Wari control. In regions devoid of constant Wari contact or supervision, people may have adapted some elements of Wari style, but not others, and modified their ideas, which resulted in the range of variability in Huamanga style ceramics. People interacted with objects in ways that worked for them. Wari allies may have manipulated their agenda to encourage subjects to feel like they belonged rather than feeling like they had been acquired.

Gosden (2005) makes an interesting argument that an object the archaeologists may identify as being "exotic," and by default very important, may have been no more important/magical/valuable than products being made by the locals because it had become local itself. For example, Gosden argues that Roman Samian ceramics were so ubiquitous in Roman Britain that they were no longer conceptualized as foreign. Although Andean scholars recognize the Huamanga ceramic style as originating in the Ayacucho region, it is not clear if Huamanga ceramics were considered exotic by locals when encountered in the provinces.[7] If, in fact, ceramics produced in Huari are not considered to be exotic and therefore important/magical/valuable objects, what could their appearance signify? Ultimately, I think the answer to this question must lie with Huamanga ceramics' production and use. If a few ceramics had been brought from the Wari heartland, some could have found their way into local ceramic workshops. Here is where Gell's notion of secondary agency of the ceramics is at work and can be observed in the archaeological record. Local potters might have taken inspiration from the ceramics, using an amalgamation of Wari and local concepts in regional ceramic production. This could have manifested in four different ways under several circumstances.

First, local populations could have been attempting to incorporate themselves into the Wari realm by using this highly visible medium found in daily and ritual practices. This would have given the ceramics the ability to interact and affect all people in the society. Interacting with and using the ceramics would have introduced the local populations to new ideas, customs, traditions, and ways of thinking and interacting. From there, the local populous could have found connections between the new ways and the old while bridging the past and the future into a path that led to Wari inclusion. This acceptance of Wari practices and iconography would have been noticed by Wari administrators and could have facilitated an easier transition into the imperial state. It could be argued that a closer fidelity, or attempted fidelity, to Wari ceramic styles would have been encouraged and aspired toward.

Second, local populations could have been attempting to resist Wari incursion and were thus subverting local messages or representations of local identity within the Wari schema (Anders 1989). As such, a closer maintenance to local notions of identity could have been located within the Wari arrangement. The potters would have manipulated Wari people (mainly administrators) by superficially following Wari "rules." Additionally, they could have been borrowing non-Wari motifs and forms found in other regions (Anders 1989).

Third, the Wari could have been incorporating local values and melding them to Wari identity as a way of proselytizing the community. In this case, if the Wari were creating the ceramics, it could have been manifested in the accurate replication of Wari notions of arrangement and iconography while erring in the reproduction of local traditions. If the local potters were to make the ceramics, then an accurate representation of local concepts would be apparent in the outcome. However, it would seem that over time, the more central local traditions would be phased out to a more Wari-dominant regiment.

Additionally, a fourth, unintentional outcome may have occurred: the potters may have been attempting to replicate the Huamanga styles, but because they lacked the specific knowledge of heartland pottery producers, they only incorporated specific elements that were recognizable and considered important. Other components may have been incorrectly replicated or left out altogether. In all of these situations, the Wari ceramics would in turn be found in all variety of households mixed with other ceramics of similar purpose. They would not be given a place of special accord or honor expected of an exotic material.

While I have presented these scenarios via the process of ceramic construction, what I want to focus on is what happened once the ceramics were made. Once completed, the ceramics would then have been experienced and interpreted by the community members. Their secondary agency would have been implicit in the social relations and their continued negotiation. In this way, while the agency of the potter, and potentially the Wari administrator, would have initiated the change, it was the ceramics that perpetuated and, arguably, shifted the relation between colonizers and colonized. The Huamanga vessels were the materialization of new and changing ideas and realities. When they were inserted into the lives and practices of the local populous, the ceramics instigated change. How the process played out depended on the intensity of the interaction and desired outcome on the part of both parties. But regardless of the result, the process of negotiation was mediated through the ceramics.

Let us return to the context of feasting. The host and feasting participants are entangled in a moment of political, religious, and social negotiation dependent on identity making and identity reinforcement. As the previously mentioned abstracted icons made their way into the everyday ceramic ware, these encapsulate moments of reinforcing cosmological ideas and their peoples' place within the Wari realm on a daily basis. Just as the process of imperial consolidation melds local social, political, and religious notions into the larger imperial whole in varying ways throughout the empire, we, too, could expect to see local influences manifesting on a day-to-day basis in objects like ceramics. Additionally, the way in which these ceramics were created allows us to move beyond observing them as solely a strategy of domination toward a multifaceted and complex process.

The variety expressed in the archaeological record, coupled with the patterned, regular placement of similar motifs on the ceramics, suggests that one of the first three explanations are more likely than the fourth option. Variation in specific motifs may reflect different local identities emerging from within the larger Wari schema. A clearer picture of Wari interactions and relationships to local peoples they encountered and/or colonized will emerge through the continued study of decorated open bowls, faceneck jars, and bowls with anthropomorphic and zoomorphic lugs.

Conclusion

Ceramics are more than utilitarian objects. They have the ability to affect social relations. Although humans create objects and bring them into existence, we are influenced by the object world we are born into and socialized within (Gosden 2005). Potters create forms and designs according to existing ideas, and in the process, they conceptualize a pot's "life." However, once a pot has been made and is in the world, interacting with other people, the potter no longer controls the object's path, and its ultimate disposition may no longer be consistent with the potter's original intention. Through these ongoing interactions with different people, the object acquires a biography of its own; no passive object has this ability. I am not, however, advocating for objects with conscious intentions or primary agency. Secondary agency cannot be achieved by the object in isolation; it demands social relations and interactions. The object cannot affect people if people do not experience it. It is through this two-sided journey that objects become imbued with meaning and history, and with power and identity.

I argue that Middle Horizon Huamanga style ceramics had the ability to act as ambassadors in place of colonists and administrators from Huari, especially in contexts that involved minimal contact with local people or in contexts of formal political, social, and ritual meetings or activities between Wari leaders and local community members (e.g., feasts). Administrators would have presented a Wari ideology that included foodways, assertions of identity, tradition, knowledge of the world, and attributes of power and prestige. In this way, the presence of ceramics in provincial contexts frees up members of the Wari heartland, and even local elites, to be useful elsewhere and/or in other ways. In many cases, the Wari ceramics became physical manifestations of social relationships. By having and using a ceramic, local people had the opportunity to replicate and participate in Wari experiences, even without Wari people being present or having to go to Huari itself. A piece of the Wari realm travels to, or is created at, another place with a different group of people.

At the same time, Huamanga ceramics constituted an avenue for local and regional ethnic groups to negotiate their identity within the Wari Empire. The category of Huamanga has taken on a "catch-all drawer" appearance to researchers because each site and group incorporated into the empire had its own unique and slightly different relationship to the empire. Consequently, the objects facilitated proselytization of their local neighbors by connecting pre-existing concepts and identities to those of the colonizing force, naturalizing the transition to a new identity under Wari leadership. Or as Anders (1986, 1989) argues, this has the potential for subverted forms of resistance by different ethnic groups.

The distribution of Huamanga ceramics in space represents not just the spread of a stylistic horizon but also a complex series of reactions, acceptances, emulations, and resistances to the Wari Empire. Gell's versatile concept of secondary agency provides a starting point for this analysis and helps us think through how materials and people interact. In this case, it provides a flexible framework that helps us make sense of ceramic style construction within the bigger and more important issue of what these ceramics did during the Middle Horizon. On a broader level, this example demonstrates how important it is to incorporate power into the discussion, even though Gell did not explicitly do so. The materiality literature often skirts around the issue of power, but all relationships are grounded on differing levels of power imbalance; if we ignore power, we downplay the importance of people, and we miss an important part of the story (Van Dyke, this volume).

In this particular study, power is integral to exploring the possible realities of the Huamanga style. Ceramic stylistic variability references power struggles that occurred simultaneously and from multiple "directions." By acknowledging that all people exercise greater or lesser degrees of power, researchers continue to move away from assumptions that nonruling peoples were passive subjects within their social and political environments, and we begin to see the ways in which top-down, bottom-up, and horizontal power engagements played out on a Wari local and regional scale. We can explore the possibilities of Wari proselytization, mutual incorporation, local resistance, and unintentional resonance for the multiple parties involved.

Gell's ideas provide a framework through which we can examine how these varying differences actually mattered to the Wari and the local communities with whom they interacted. A Gellian approach forces us to look at the question from the perspective of potential relationships between people and objects not only on the small scale of a site but also at a scale that stretches across the expanse of an empire. It allows us to avoid getting lost in traits and typologies and directs us back to what matters most—people: how they lived and who they thought they were.

I believe this approach works well for several reasons. It focuses on the people and the objects simultaneously and forces us to move back and forth between each testing how different possibilities would have played out for each party. It can work at a multiscalar level. The theory could

be applied to virtually any "category" of object uncovered by archaeological endeavors; and finally, it takes what is valuable about typological and stylistic approaches and moves beyond their shortcomings. This
agency-based analysis is possible because scholars have been collecting
Wari materials over a good expanse of time and distance. Additionally, I
had access to an extraordinarily large amount of ceramic data at the site of
Conchopata. Both these factors may impact the success of applying this
method to other contexts. Nonetheless, a Gellian perspective is likely to
be quite useful for a range of archaeological questions. It does not require
drastic changes in focus. Rather, it is built upon just a small shift in perspective, and it offers a useful tool for "thinking through" relationships
among past peoples and materials. Gell's ideas can be used to produce
new insights, reflections, and, ultimately understanding of the possibilities for a moment in time.

Acknowledgments

This chapter evolved from my MA thesis and could not have been ac-
complished without the foundational support, wisdom, and provision of
data by William H. Isbell. The ideas herein have been nurtured through
various discussions with the "Agency Club," both in seminar and outside,
over the past few years. Thank you to Şule Can, Tanya Chiykowski, Rui
Gomes Coelho, Erina Gruner, Jessica Santos López, and Halona Young-
Wolfe. An additional thank you to all of you who read drafts of this chapter
and provided excellent advice. I would also like to thank Anne Hull who
digitized and drafted my drawings. A special thank you to Ruth M. Van
Dyke, who not only brought us together and helmed this volume, but who
put a tremendous amount of care, time, and hard work into each of our
projects and lives. A final thank you to my family—your love and support
allowed me to pursue my dreams. Any errors or problems in the chapter
are my own.

Notes

1. I conflate Gell's notion of artist with what archaeologists refer to as *maker* or
producer. I believe the idea of the role to be applicable to those who created and made
objects of the past that may or may not have been considered "art." However, I will
continue to refer to the term *artist*, as it is what Gell used (in addition to the terms
maker or producer).

2. The organization of production is unresolved; Isbell (2007) proposed production by polygynous wives in a household context, while others have argued for ceramic workshops (Cook 1987; Isbell and Cook 1987).

3. Following Isbell (2001:457), Wari is the culture, and Huari is the capital site.

4. A note of disclaimer: I do not in any way see ceramic styles as a bounded, static set of traits. I acknowledge the ability of producers of ceramics to be creative and able to renegotiate or change over time. It is a process that is affected by other societal influences and the interactions that pre-existing ceramics have with members of the society. However, all archaeologists must create some kind of arbitrary classifications in order to be able to say anything about the past. It is recognized that these boundaries are arbitrary and may or may not have been relevant or significant to the people in the past who produced said artifacts. All we can do is try to identify the patterns of concepts/ideas/notions/point of views imbued within the material culture. In this way, we are moving from the stance of object mutability to a collective ceramic style's ability to be fluid and changing, depending on the process and outcomes of social interactions. As Stahl states, "genealogies of practices are not necessarily built upon exact replication, but on the reassembly of things that have historical and meaningful referents" (2008:185). In order to observe ceramic attribute commonalities in an unrestricted manner, I employ concepts of agency to the debate on Huamanga and Huamanga-like ceramics.

5. Wari ceramics consist of mainly open bowl forms (straight, flaring, and curved) and restricted jar forms. Additional less common forms include keros, tumblers, lyre cups, urns, modeled figures, bottles, and canteens. Vessels are slipped (buff, orange, cream, red, or black) and unslipped. Most of the decoration is painted polychrome utilizing black, white, orange, red, grey, and purple. Finishes to slipped and painted vessels are either matte or burnished depending on style. Incised decoration occurs less frequently across Wari ceramic traditions, and is commonly attributed to the Black-Incised style. Decoration on the ceramics consists of geometric and figural motifs.

6. The problems mentioned previously plague the Huamanga ceramic style. For example, Huamanga has been described as secular Viñaque (Anders 1989) and regular (not fancy) Chakipampa (Groleau 2005); both authors cite Menzel (1964). It additionally has been described as Atarco, regular Viñaque, Pinilla, and Q'osqopa (Owen 2007), although other ceramics not considered to be Huamanga belong to those categories as well. In the Cuzco region, the style of Araway/Arahuay has been argued to be Huamanga (Glowacki 1996).

7. The style is attributed to the heartland region, however it is unclear where many Huamanga and Huamanga-like ceramics were manufactured. Arguably, Huamanga-like ceramics were locally produced; sourcing studies would be a fruitful avenue for future research.

References Cited

Alberti, Benjamin. 2012. Cut, Pinch and Pierce: Image as Practice Among the Early Formative La Candelaria, First Millennium AD, Northwest Argentina. In *Picture This! The Materiality of the Perceptible*, edited by I. M. Back Danielsson, F. Fahlander, and Y. Sjöstrand, pp. 23–98. Stockholm University, Stockholm.

Alberti, Benjamin, and Yvonne Marshall. 2009. Animating Archaeology: Local Theories and Conceptually Open-ended Methodologies. *Cambridge Archaeological Journal* 19(3):344–356.

Anders, Martha B. 1986. Dual Organization and Calendars Inferred from the Planned Site of Azángaro: Wari Administrative Strategies (Volumes 1–3). Unpublished PhD dissertation, Department of Anthropology, Cornell University, Ithaca.

———. 1989. Wamanga Pottery: Symbolic Resistance and Subversion in Middle Horizon Epoch 2 Ceramics from the Planned Wari Site of Azángaro (Ayacucho, Peru). In *Cultures in Conflict: Current Archaeological Perspectives*, edited by Diana Claire Tkaczuk and Brian C. Vivian, pp. 7–18, University of Calgary, Calgary.

Bauer, Brian. 2002. *Las antiguas tradiciones alfareras de la región del Cuzco*. Centro de Estudios Rurales Andinos "Bartolemé de Las Casas," Cuzco.

Benavides Calle, Mario. 1965. Estudio de la Cerámica Decorada de Qonchopata. MA thesis, Departamento de Antropología, Universidad Nacional de San Cristobal de Huamanga, Ayacucho Perú.

Cook, Anita G. 1987. The Middle Horizon Ceramic Offerings from Conchopata. *Ñawpa Pacha* 22–23:49–90.

———. 2012. The Coming of the Staff Deity. In *Wari: Lords of the Ancient Andes*, edited by Susan E. Bergh, pp. 103–121. Thames and Hudson, New York.

Cook, Anita G., and Mary Glowacki. 2003. Pots, Politics, and Power: Huari Ceramic Assemblages and Imperial Administration. In *The Archaeology and Politics of Food and Feasting in Early States and Empires*, edited by Tamara L. Bray, pp. 173–202. Kluwer Academic/Plenum, New York.

Gell, Alfred. 1992. The Technology of Enchantment and the Enchantment of Technology. In *Anthropology, Art, and Aesthetics*, edited by J. Coote and A. Shelton, pp. 40–67. Clarendon Press, Oxford.

———. 1998. *Art and Agency: An Anthropological Theory*. Clarendon Press, Oxford.

Glowacki, Mary. 1996. The Wari Occupation of the Southern Highlands of Peru: A Ceramic Perspective from the Site of Pikillacta. Unpublished PhD dissertation, Department of Anthropology, Brandeis University, Waltham, Massachusetts.

Glowacki, Mary, and Michael Malpass. 2003. Water, Huacas, and Ancestor Worship: Traces of a Sacred Wari Landscape. *Latin American Antiquity* 14(4):431–448.

Gosden, Chris. 2005. What Do Objects Want? *Journal of Archaeological Method and Theory* 12(3):193–211.

Groleau, Amy B. 2005. House-Keeping and House-Leaving: A Case Study of Oversize Ceramic Offering and Modes of Abandonment from Middle Horizon, Peru. Unpublished MA thesis, Department of Anthropology, Binghamton University, New York.

———. 2009. Special Finds: Locating Animism in the Archaeological Record. *Cambridge Archaeological Journal* 19(3):398–406.

Herring, Adam. 2010. Shimmering Foundation: The Twelve-Angled Stone of Inca Cusco. *Chicago Journals* 31(1):60–105.

Ingold, Tim. 2007. Materials against Materiality. *Archaeological Dialogues* 14(1):1–38.

Isbell, William H. 1977. *The Rural Foundations for Urbanism: Economic and Stylistic Interaction between Rural and Urban Communities in Eighth-Century Peru*. Illinois Studies in Anthropology No. 10. University of Illinois Press, Urbana.

———. 2001. Reflexiones Finales. In *Huari y Tiwanaku: Modelos vs. Evidencias, segunda parte*, edited by Peter Kaulicke and William H. Isbell, pp. 455–479. Boletín

de Arqueología PUCP, vol. 5, Departamento de Humanidades, Especialidad de Arqueología, Pontificia Universidad Católica del Perú, Lima.
——. 2007. A Community of Potters or Multicrafting Wives of Polygynous Lords? In *Craft Production in Complex Societies: Multicraft and Producer Perspectives*, edited by Izumi Shimada, pp. 68–96. University of Utah, Salt Lake City.
Isbell, William H., and Anita G. Cook. 1987. Ideological Origins of an Andean Conquest State. *Archaeology* 40(4):27–33.
Isbell, William H., and Patricia J. Knobloch. 2006. Missing Links, Imaginary Links: Staff God Imagery in the South Andean Past. In *Andean Archaeology Vol. III: North and South*, edited by William H. Isbell and Helaine Silverman, pp. 307–351. Springer, New York.
——. 2009. SAIS—The Origin, Development and Dating of Tiahuanaco-Huari Iconography. In *Tiwanaku: Papers from the 2005 Mayer Center Symposium at the Denver Art Museum*, edited by M. Young-Sanchez, pp. 163–210. Frederick and Jan Mayer Center for pre-Columbian and Spanish Colonial Art at the Denver Art Museum, Denver.
Lumbreras, Luis. 1974. *Las fundaciones de Huamanga: Hacia una prehistoria de Ayacucho*. Club Huamanga, Lima.
McEwan, Gordon F., and Patrick Ryan Williams. 2012. The Wari Built Environment: Landscape and Architecture of Empire. In *Wari: Lords of the Ancient Andes*, edited by Susan E. Bergh, pp. 103–112. Thames and Hudson, New York.
Menzel, Dorothy. 1964. Style and Time in the Middle Horizon. *Ñawpa Pacha* 2:1–106.
——. 1968. *La Cultura Huari*. Compania de Seguros y Reseguros Peruano-Suiza S. A., Lima.
Mills, Barbara J., and T. J. Ferguson. 2008. Animate Objects: Shell Trumpets and Ritual Networks in the Greater Southwest. *Journal of Archaeological Method and Theory* 15:338–361.
Nash, Donna J. 2012. The Art of Feasting: Building an Empire with Food and Drink. In *Wari: Lords of the Ancient Andes*, edited by Susan E. Bergh, pp. 82–100. Thames and Hudson, New York.
Owen, Bruce. 2007. Rural Wari Far From the Heartland: Huamanga Ceramics from Beringa, Majes Valley, Peru. *Andean Past* 8:287–373.
——. 2010. Wari in the Majes-Camana Valley: A Different Kind of Horizon. In *Beyond Wari Walls: Regional Perspectives on Middle Horizon Peru*, edited by Justin Jennings, pp. 233–254. University of New Mexico Press, Albuquerque.
Schreiber, Katharina. 1992. *Wari Imperialism in Middle Horizon Peru*. Anthropological Papers, Museum of Anthropology, University of Michigan No. 87. Museum of Anthropology, University of Michigan, Ann Arbor.
——. 2001. The Wari Empire of Middle Horizon Peru: The Epistemological Challenge of Documenting an Empire Without Documentary Evidence. In *Empires: Perspectives from Archaeology and History*, edited by Susan E. Alcock, Terrance N. D'Altroy, Kathleen D. Morrison, and Carla M. Sinopoli, pp. 70–92. Cambridge University Press, Cambridge.
——. 2012 The Rise of an Andean Empire. In *Wari: Lords of the Ancient Andes*, edited by Susan E. Bergh, pp. 31–44. Thames and Hudson, New York.
Stahl, Ann B. 2008. Dogs, Pythons, Pots, and Beads: The Dynamics of Shrines and Sacrificial Practices in Banda, Gahna, 1400–1900 CE. In *Memory Work: Archaeologies*

of Material Practices, edited by Barbara J. Mills and William H. Walker, pp. 159–186. School of Advanced Research Press, Santa Fe.

Viveiros de Castro, Eduardo. 1998. Cosmological Deixis and Amerindian Perspectivism. *Journal of the Royal Anthropological Institute* 4(3):469–488.

———. 2004. Exchanging Perspectives: The Transformation of Objects into Subjects in Amerindian Ontologies. *Common Knowledge* 10(3):463–484.

Weismantel, Mary. 2013. Inhuman Eyes: Looking at Chavín de Huantar. In *Relational Archaeologies: Humans/Animals/Things*, edited by Christopher Watts, pp. 1–20. Routledge, London and New York.

Williams, Patrick Ryan, and Donna J. Nash. 2006. Sighting the Apu: A GIS Analysis of Wari Imperialism and the Worship of Mountain Peaks. *World Archaeology* 38(3):455–468.

Zedeño, Maria Nieves. 2008. Bundled Worlds: The Roles and Interactions of Complex Objects on the North American Plains. *Journal of Archaeological Method and Theory* 15:362–378.

CHAPTER SEVEN

The Work They Do

Phenomenology and Monumentality in the Late Archaic of Peru

Halona Young-Wolfe

Because archaeology is a discipline with a specific and unique concern with the material, it makes sense that archaeologists would seek to utilize theories of materiality. However, a problem emerges in many such analyses in that the field lacks a strong set of methods to incorporate the theories of materiality and agency into archaeological analysis. How do archaeologists connect theories of materiality to the archaeological record?

Phenomenology shares a concern with the ways that people and things are together in the world and so offers a promising approach to materiality in archaeology. However, the way in which phenomenology has been mobilized thus far has received a robust critique from within the discipline. This chapter proposes a different approach to phenomenology in archaeology. Linking Martin Heidegger's phenomenology with concepts from ecological psychology addresses key concerns about phenomenology as a method for archaeology. This fusion of ideas allows archaeologists to refocus on the way that people and things are together in the world and to direct archaeological questions away from modern concerns and experiences toward issues of context, *praxis*, and experience in the past. This method enables a fresh look at the appearance of some of the first monumental architecture in the Americas, which date to the Late Archaic period on the North Central Coast of Peru. Archaeological studies of these sites have portrayed the appearance of monumentality as diagnostic of a sharp break with previous social, political, and economic systems, as well as a material sign of emergent social hierarchy and political authority. What these studies have lacked is a concern for the process and human experience of this newly altered landscape. Through attention to a distinctive and understudied construction technique, the use of bagged construction fill known as *shicra*, the construction of these monuments can be understood as organic to the lifeworld of the ancient Andean peoples. Thus, the work that monuments did can be understood as conveying continuity and fidelity with their past, even as they changed the shape of their future.

Materiality, Phenomenology, and Archaeology

Materiality and Archaeology

Calls to attend to material, though they are not limited to archaeology, make particular sense from within the discipline. Bjørnar Olsen (2010:22) characterizes archaeology as "the foremost discipline of things." Although archaeology has a clear affinity with material, by applying theories of materiality and agency to the questions of archaeology important issues arise. The intellectual heritage of the Cartesian separation of subject and object and the body and mind—a uniquely Western and modern approach to the world—strongly impacts the ways in which anthropologists and archaeologists understand and encounter the physical world (Boivin 2008; Ingold 2007; Thomas 1996). Because of the primacy of the subject that is implicit in the Cartesian separation of subject and object, signified and sign, anthropological and archaeological discourse on materiality has largely produced interpretations focused on meaning that reiterates and replicates the privileged position of the ideal over the material (Ingold 2005:123; Küchler 2005:207; Miller 2005:30; Olsen 2010:16; Pinney 2005:257; Smith 2003:67). This focus on representation and meaning has made it challenging to think about quotidian objects and their roles in past lifeworlds (Hurcombe 2007:536; Olsen 2010:59–60). Without the ability to see objects as objects, rather than as signs, it is impossible to see objects as anything other than communicators of human thought and intent (Boivin 2008:21; Olsen 2010:13). The focus on the ideal also disguises process and action in favor of the idea of an outcome or finished product. This is another hallmark of our modes of modern Western thought that may not be well aligned with the experience of past peoples, and it deflects attention away from the important processes of becoming and embodiment (Ingold 2000:198).

Archaeology, then, is a discipline that has a unique and specific concern with objects and how objects and people act upon and interact with one another. This makes questions about materiality and agency particularly important for the field. However, in approaching materiality and agency, there is a gap between the high-level theory gleaned from figures such as Alfred Gell (1998) or Bruno Latour (2005) and concrete methods needed to articulate these ideas with the archaeological record. The question remains: Is there a way to attend to the material world in a way that enhances our understanding of the lives of people of the past?

Phenomenology and Archaeology

For many, phenomenology has been a way to bridge the link between materiality theory and archaeological practice. As Joanna Brück (2005:46) states, "for a discipline which argues for the social, cultural, and ontological centrality of objects to the human species, phenomenological approaches clearly provide an antidote to abstract models which prioritize the role of the mind in human cognition." Particularly, many archaeologists have deployed phenomenology in investigations of the British landscape. The engagement of archaeologists, such as Tilley (1994, 2004) and Cummings and Whittle (2004), with phenomenology arises out of postprocessual archaeology (Olsen 2010:27) and specifically postprocessual critiques of processual approaches to landscape archaeology (Fleming 2006:268). In particular, postprocessual archaeologists look to phenomenology to return people to the investigation of landscapes, to illuminate through embodied practice the ways in which meaning is attached to landscapes, and to dissolve subject/object dualisms (Barrett and Ko 2009; Brück 2005; Fleming 1999, 2006). But according to critics, they have attained questionable success in meeting these goals.

Critics of the way phenomenology has been deployed in landscape archaeology ask a basic question: Would the relationships, connections, and meanings identified through contemporary encounters with the landscape have in fact been significant to people in the past (Barrett and Ko 2009; Brück 2005; Fleming 2006)? Brück (2005:51) characterizes this issue as the difference between association and causation. For example, what appears to a contemporary archaeologist to be meaningful choices about where and how to place monuments may reflect the differential availability of materials, differential preservation (Fleming 1999:120), or "esoteric" desires that would leave no physical trace for future observers (Barrett and Ko 2009:283). As Brück (2005:52) states, "the problem, then, is that relationships claimed are not always demonstrated or supported adequately." And, as Fleming (2006:273) has pointed out, it is not clear how in fact a phenomenological approach would or could provide proof or support.

Additionally, as Tim Ingold notes in his review of Christopher Tilley's 2004 book, *The Materiality of Stone: Explorations in Landscape Phenomenology*, projects intending to focus on embodied experience frequently evolve into discussions that are primarily about thought and belief. Ingold (2005:123) casts this as a failure of the phenomenological project and locates the cause in the failure of archaeologists to break with the traditional approach of material culture studies that "treats the physical world as a pool of metaphorical resources for the expression of social or cosmological principles." By continuing to privilege the ideal over the material, archaeologists fail to realize the potential of phenomenology.

Phenomenology, then, has been a problematic method with which to approach archaeological investigation into issues of materiality. Yet phenomenology promises to accomplish many of the goals of such an investigation. The question becomes, then, is there a different way to mobilize phenomenology that will overcome these difficulties and help bridge the divide between a theory of materiality and the practice of archaeology? In particular, several characteristics of Martin Heidegger's phenomenology make it well suited to archaeological investigations; this point that has been made by John Barrett and Ilhong Ko (2009), Tim Ingold (2000), and Bjørnar Olsen (2010).

Julian Thomas (1996) has made a careful exploration of Heidegger's ideas and their utility for archaeology in *Time, Culture and Identity*. Thomas (1996:29) makes a particularly strong argument against the "self-defeating" project of modernist archaeology. For him, Heidegger's theories allow a recognition and analysis of the unique human characteristics of "temporality and the use of cultural knowledge" through the interconnectedness of persons and things (Thomas 1996:237). One of Thomas's main concerns is to propose an antidote for arguments that meaning is absent from archaeological materials because meaning is created and housed in individual human minds. His approach represents a thoughtful incorporation of Heidegger's work into archaeological interpretation, yet his main objects of analysis—henges, burials, and hoards of ritual objects—fall outside the realm of day-to-day life.

However, as Olsen (2010:87) has noted, Heidegger has a "detailed concern with mundane things in their everyday uses [and] with equipment and their referential assignment." It is precisely this "detailed concern" that makes Heidegger's phenomenology uniquely useful for the discipline. This is especially true of studies that focus on quotidian objects, a typically understudied subject in many regional archaeologies that is now seeing increased interest (see Chiykowski this volume; Fullen this volume).

Heidegger's Phenomenology: Equipment and Dasein

Heidegger (1982 [1954]:21) explicitly defined his phenomenology as the scientific method by which *Dasein* (the being of humans) could be understood. The method is based on Heidegger's understanding of Dasein as a being whose existence is inextricably intertwined with things. Heidegger does not treat things as symbols or as projections of human agency. Instead, he sees the ontological state of Dasein as essentially linked to quotidian things.

> [Dasein] never finds itself otherwise than in the things themselves, and in fact in those things that daily surround it. It *finds itself* primarily and constantly *in things* because, tending them, distressed by them, it always in some way or other rests in things. Each one of us is what he pursues and cares for. In everyday terms, we understand ourselves and our existence by way of the activities we pursue and the things we take care of. (Heidegger 1982 [1954]:159, emphasis in original)

Because Dasein finds itself through daily interaction with things, the method of phenomenology does not proceed from theory but instead proceeds from action. Christopher Macann (1993:73) describes this approach by saying "according to Heidegger, entities simply are not there, first and foremost, for theoretical inspection and examination. Rather, we do things with them, pick them up, discard them, manipulate them, [and] put them to use."

The things with which Dasein is concerned, which Heidegger calls *equipment* (or tools), exist in a totality of relationships. The functionality of equipment is essential to understanding its nature. "Each individual piece of equipment . . . has its immanent reference to that *for which* it is what it is" (Heidegger 1982 [1954]:163, emphasis in original). Equipment is always part of an "equipmental" whole, and so stands in relation to other equipment. The functionality of each piece of equipment must be understood not as a property attached to a thing, nor as a property that only exists in relations between things, but rather as essential to each piece of equipment; that which makes the thing what it is.

It is into this world of equipment, the *umwelt*, that Dasein is thrown. These are the terms of a Dasein's existence, which is *being-in-the-world* (Heidegger 1982 [1954]:164). And it is through concrete activity, by seeking, touching, looking, and doing that Dasein is thrown into the world and finds itself in the world (Aho 2009:13). Thus, the mundane functionality of everyday objects and active interaction with these objects is inextricably linked to the ontological status of all people.

The potential of a phenomenological approach based on Heidegger's concern for equipment is promising for an archaeological approach to materiality because it insists upon the interconnectedness of people and objects and because it attends to the concrete, functional aspect of people and things in the world. Heidegger's emphasis on the essentially functional nature of equipment is particularly useful for thinking about everyday objects; the undecorated pots, unretouched flakes, and various utilitarian objects that comprise the majority of artifacts at most archaeological sites are frequently neglected in discussions of materiality.

Yet, to mobilize Heidegger's phenomenology in an archaeological context, we must be mindful of the important differences between Heidegger's ontological project and archaeology's historical project. Heidegger's Dasein, though sometimes interpreted as a subject, in fact should be understood as a way of being (Aho 2009:12). Although Heidegger's phenomenology is focused on the everyday life of humans, Dasein does not represent an embodied human, and Heidegger is not concerned with a specific subject, world, or experience. In exploring Heidegger's treatment of the body and the many criticisms of that treatment, Kevin Aho (2009:4) notes that toward the end of his career, Heidegger himself acknowledged the challenges of dealing with the body, calling the "body problem" one of the most difficult to think about.

Therefore, to mobilize Heidegger's phenomenology for archaeological investigations, we must find a way to account for the physical human being as an integral part of the specific past we wish to investigate. Webb Keane (2005) points toward an interesting possibility in how to solve this "body problem" when he discusses how objects are oriented toward the future. For example, through its material form, a chair creates possibilities of how it may interact with people. "What interests us as embodied actors, rather than, say, spectators, is the chair's instigation (by virtue of its form) to certain sorts of action—and thus, its futurity" (Keane 2005:194).

This ability of objects to act and influence the action of people through their physical properties is developed by psychologist James Gibson in his ecological psychology. For Gibson, this invitation of the chair to sit, which matches with the human ability to bend at the middle and at the knees and so conform to the chair by sitting, is an affordance. Timothy Ingold (2000, 2007) has explored the utility of Gibson's theories for archaeology and materiality, and Olsen (2010:146) notes that Gibson's idea that objects "offer their properties and ready-to-actness in a direct, unmediated way" is a helpful way of approaching objects for archaeology. The concept of affordances echoes the functionality of equipment that is so central to Heidegger's philosophy and also reasserts the presence of the embodied human in interaction with things, which is unaccounted for in Heidegger's approach. When taken together, these theories can dissolve the subject-object divide, recast the human as an inseparable component of the physical environment, and transform objects from outcomes of internal human processes to active and constituent parts of cognitive systems that extend across the skin barrier. In the process, objects and bodies recover their material properties and the physical becomes a powerful place from which to investigate everyday life. When applied to archaeology, ecological psychology offers tools to use in thinking about materiality and phenomenology that open new possibilities for understanding how people and objects are in the world together.

Ecological Psychology

Ecological psychology contends that perception does not happen internally within an animal; it is a process in a system that consists of an animal and its environment (Michaels and Carello 1981:1). It is built on James Gibson's theory of direct perception, a very different model of how animals receive, process, and react to information than the more familiar representational models. The impact of the difference between these approaches is clearly demonstrated with an example from developmental psychology.

In developmental psychology, the study of very young children has tended to focus on the timing of the development of specific mental concepts. However, reframing classic experiments to attend to the body has shown that looking at behavior from an embodied perspective shows how "intelligence emerges in the interaction of an organism with an environment" (Smith 2005:278). This is demonstrated by Linda B. Smith and her colleagues, who have reframed classic experiments from developmental psychology to attend to bodily action rather than mental concepts. One such experiment is Piaget's "A not B" experiment. This test has been used to determine when infants develop an *object concept*, that is, the belief that objects persist even when we are not interacting with them (Smith 2005:280). The object concept is a classic example of representational thinking, and Piaget's simple test has been used to determine that this mental concept develops at about the age of twelve months.

The experiment works by showing an infant a toy and then hiding the toy under a lid or box at location A. After a short delay, the infant is allowed to reach for the toy. The procedure is repeated several times, but then the experimenter changes the hiding place of the toy to location B. It is important to remember that in all the trials, the infant *watches* as the toy is hidden. After the same short delay, the infant is once again allowed to reach for the toy. Infants between eight and ten months old make the classic A not B error; they reach for the toy at the old location, A, even though they have seen it being placed at B. Typically, after the age of twelve months, the infants are able to successfully reach for the toy at location B (Smith 2005:280–281). The classic explanation has been that the younger children fail because they lack an object concept; they do not understand that the toy exists independent of their own perception.

However, Smith and her colleagues (2005) have shown that simple changes to the test change the results and allow the younger infants to correctly locate the toy at location B. For example, shifting the baby's posture (from sitting to standing) as it watches the toy being placed at location B or adding weights to the baby's arms change the experimental results. The researchers feel these findings support a dynamic, embodied model of cognition. They argue that it is not the lack of an object concept but rather the repeated bodily experience of reaching for the toy at location A that causes the younger children to make the error. Introducing physical changes, such as shifts in posture, interrupts the habitual reaching response and allows the infant to reach for the toy at location B. Smith (2005:285) explains "the A not B error cannot be explained by a constant (or its lack) in the head, but it can be explained by the processes that bound cognition through the body to the here and now world. Again, a fixed belief—an object concept—seems to have little to do with it."

From the perspective of ecological psychology, the body is an important component of the cognitive system, and bodily interactions with things can be understood as a part of cognition. The relation between the body and objects is mediated by *affordances*. "An affordance is neither an objective property nor a subjective property; or it is both if you like. An affordance cuts across the dichotomy of subjective-objective and helps us to understand its inadequacy. It is equally a fact of the environment and a fact of behavior. It is both physical and psychical, yet neither. An affordance points both ways, to the environment and to the observer" (Gibson 1979:129).

Affordances are not located in either the animal or the environment because they are *relations* between the abilities of a specific animal and the features of a specific environment (Chemero 2009:145). The innovative aspect of affordances is that, according to Gibson's view, it is affordances that are perceived rather than objects. As Michaels and Carello (1981:42) explain, "we would say that humans do not perceive chairs, pencils and doughnuts; they perceive places to sit, objects with which to write, and things to eat. To say that affordances are perceived means that information specifying these affordances is available in the stimulation and can be detected by a properly attuned perceptual system. To detect affordances is, quite simply, to detect meaning."

The concepts of information and meaning used here are specific. Information in ecological psychology is always information about an environment and for an animal (Michaels and Carello 1981:38). The information is not independent from the animal. This is because the information consists of the physical characteristics of the environment, in interaction with the specific body of the animal, and it is obtained through the information-seeking behavior of the animal. Michaels and Carello (1981:46) use meaning to refer to what the animal can do with the object. Meaning, in this discussion, should not be understood to be about concepts, ideals, or symbols. Meaning is always information about action.

Living creatures, then, interact with their environment through the affordances that facilitate or resist their action, and accordingly, they actively attend to their physical environment rather than acting simply as passive receptors (Barrett 2011:106). In doing this, they shape their environment, but as they build upon their interactions, they are shaped by the environment. This sense of the body is very different from the understanding of embodiment that is commonly found in the archaeological literature that is inspired by Cartesian ideals, which sees a natural state for humans as "unmixed" with their environment (Olsen 2010:36). Smith and colleagues' reworking of the A not B experiment is just one example from a corpus of experiments, human and animal, that support the importance of the interaction of the body and the world in predicting and explaining behavior. Many of these results have originally been explained through computational and representational models of cognition. Previous trials gave little attention to variables that included the body. But when the interaction of the body and the environment is taken as important, experimental results emerge that support theories of ecological psychology and embodied cognition.

These results highlight the false separation of animal from environment. If the cognitive process takes place within an interaction of animal and environment, there is no justification for dividing the organism from the environment. Here is where the Cartesian subject-object divide dissolves, as Barrett (2011:199) clearly states.

> If we think of cognition as an active process, and "mind" as something animals do rather than something they "have," then questions about whether "minds" are things inside the head, or things that can exist outside them, don't really make much sense. The metaphor of containment that we are using to think about these things—that a mind is a thing inside a person and distinct from the outside world—begins to break down.

The insistence of ecological psychology that meaning rests in the relationship between animal and environment, and that meaning is about action, aligns closely with Heidegger's phenomenology. Heidegger argued that our everyday actions are a form of "smooth coping." That is, everyday action is nonrepresentational (Barrett 2011:149). For example, when we use a familiar tool, we do not need to create a mental image of ourselves "as a subjective mind directed toward . . . an object" (Barrett 2011:149). Heidegger would describe the tool as being "ready to hand." It is through the interaction with equipment, the "concern" with things, and the inextricable linkage of Dasein and the world that Heidegger believes humans are constituted in the world.

Cognitive Systems, Equipment, and the Archaeological Record

If we use the tenets of ecological psychology as a tool to think about how to deploy phenomenology in archaeology, we can envision a phenomenology that can avoid the pitfalls and problems that have emerged in the past. This perspective truly integrates people in their environment as constituent parts of a dynamic cognitive system that exists in the world. Within this system, the physical properties of the constituent parts, the "thingness of things" and the physicality of the body, are restored to the center of attention. And, if we accept the idea that much daily behavior is based on nonrepresentational cognition, it becomes much more appropriate to step away from searches for symbolic meaning in dealing with artifacts. The more appropriate question becomes "how did these objects form part of a system that included human bodies and behavior?"

The importance of this approach is not that it creates humans and things as separate but equal kinds of beings. Instead, it places, first and foremost, people and things together in a system. One does not proceed the other or in fact exist independent of the other. Rather, both develop together and through one another. As Thomas (1996) points out, this is a key difference between an approach that utilizes Heidegger's phenomenology and one that relies on systems theory. We cannot attempt to discuss "persons" or "things" because each always already implicates the other. Thus, this approach not only enables but requires an analysis that places people and objects together at the center of every archaeological question.

Linking concepts of ecological psychology with Heidegger's phenomenology, then, offers new ways to think through problems that confront the archaeologist who wants to engage with materiality. There is a strong resonance between this approach and Ingold's idea of a *taskscape*. Ingold (2000:195) envisions the taskscape as the interlocking array of concrete tasks that are carried out in daily life. The taskscape is embedded within a specific environment that is embodied, temporal, and inextricably social, and the tasks are "the constitutive acts of dwelling" (Ingold 2000:195). For Ingold (2000:198) the landscape can be understood as "the taskscape in its embodied form: a pattern of activities collapsed into an array of features." What an archaeology of any given landscape can hope to recover, then, is an echo of the taskscape that produced it. Rather than an investigation of a discreet, abstract segment created through modern theorizing, such as a study of economic systems or ritual activity, we can instead expect that a study based on this approach will evoke the system as a whole and produce an understanding that is much more concerned with daily, bodily experience.

Shicra and Late Archaic Monumental Architecture of North Central Coast Peru

Archaeological sites on the North Central Coast of Peru are distinguished by the emergence of monumental architecture in the Late Archaic period. Research by Peruvian and North American scholars has largely focused on the investigation and interpretation of the monuments and with very good reason. These are among the earliest known monumental constructions in the Americas. The following case study offers a new approach to exploring the monuments. I focus not on their uniqueness, large scale, and disruptive appearance on the landscape but rather on a mundane and small-scale construction technique—the use of bagged fill known as shicra. By using the theoretical basis presented in the preceding discussion to explore the use of shicra, we can see how the monuments of this area can be understood as an integrated and organic part of their particular dynamic system.

The Late Archaic, North Central Coast of Peru

Hundreds of archaeological sites from throughout Andean prehistory are located in the four adjacent river valleys and associated desert coastal areas of the North Central Coast of Peru (Figure 7.1). Dozens of these sites have occupations dating between 3000 and 1800 BC (Late Archaic period) and feature some of the oldest monumental architecture in South America. The appearance of the monumental architecture in this preceramic society has gained a great deal of attention from researchers because there was only a very limited architectural tradition in this area before the appearance of the monuments; no remains of any buildings other than small domestic units have been identified, and these are primarily constructed of perishable materials.

Additionally, there has been much debate about the economic basis of the society and its ability to support large populations with "complex" social structures. The limited evidence for intensive agriculture led Michael Moseley to develop a theory, which is published in his 1975 title *Maritime Foundations of Andean Civilization*. Moseley (1975) proposed that the Andes were the site of the wholly unique development of a pristine civilization supported from maritime resources rather than agriculture. These sites are widely characterized as the birthplace of Andean civilization and have inspired decades of archaeological research and lively debate (Feldman 1985; Haas and Creamer 2006; Haas et al. 2004; Moseley 1975; Pozorski and Pozorski 2008; Shady 2004). These debates have primarily been concerned with fitting this Late Archaic culture into a system of sociopolitical typologies primarily based on universal evolutionary models. The monumental architecture has most frequently been evaluated in quantitative terms and the idea that measuring the energy invested in the monuments could define the nature of the corporate entity that directed the building programs (Vega-Centeno 2010). Thus, the prevailing research agendas have been to quantify, define, and label. The result

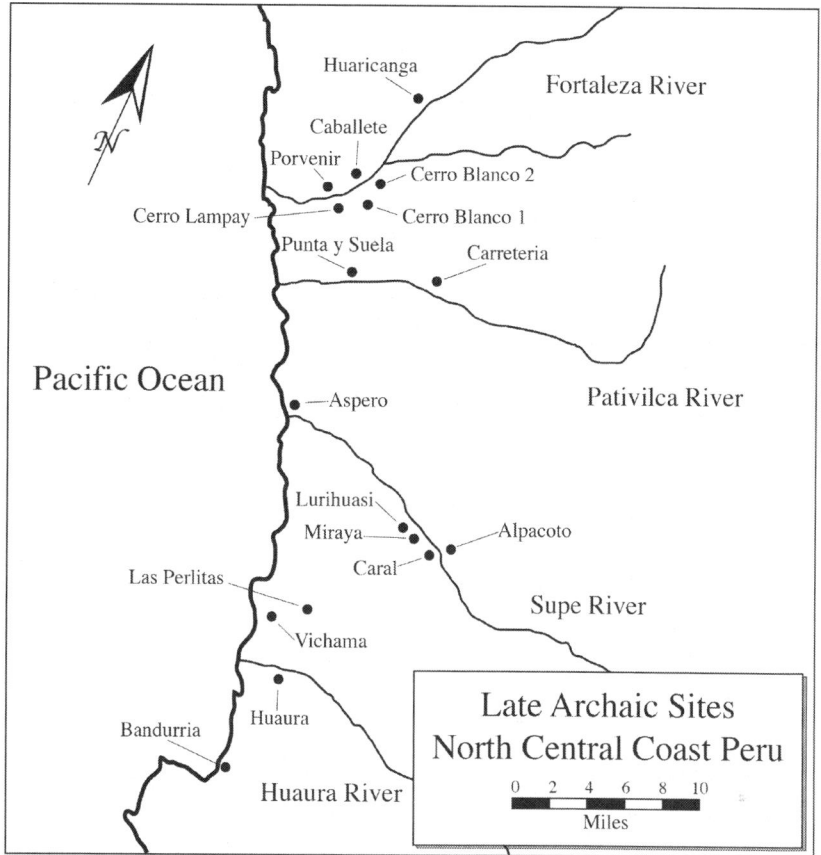

Figure 7.1. Late Archaic sites on the North Coast of Peru. Drafted by Anne Hull.

is an understanding that has much to say about modern ideas of power and hierarchy, but little to say about the lives of the people who lived with these striking monuments.

The Fortaleza, Pativilca, Supe, and Huaura valleys have more than thirty Late Archaic sites featuring monumental architecture; each of these sites has between one and seven large-scale platform mounds (Haas and Creamer 2012:296). There are clear similarities in construction materials, methods, and design within and between valleys. The typical architecture was built over multiple episodes and consists of terraced platform mounds, or truncated pyramids, with enclosed rooms on the top level. The mounds are typically symmetrical and feature a central staircase that connects with a circular sunken plaza located in front of the monument

Figure 7.2. Pyramidal Mayor at Caral, Supe Valley, showing the typical symmetrical construction, central staircase, and circular sunken plaza. Multiple construction phases are also evident. Photo by Halona Young-Wolfe.

([Figure 7.2] Haas and Creamer 2012; Pozorski and Pozorski 2012; Quilter 1991).

However, Jeffrey Quilter (1991:402) points out that shicra, one of "the most diagnostic features" of Late Archaic monumental construction on the coast, is much more modest and less impressive to the casual visitor (Figure 7.3). Shicra are looped net bags made of locally growing sedges, primarily *Scirpus totora*, that are filled with stone. A typical shicra, as described by Rafael Vega-Centeno (2005:97) from his excavations at Cerro Lampay in the Fortaleza Valley, can hold between twenty and twenty-five kilograms of medium- to small-sized stones. They are used to fill rooms and between walls during construction and renovation and have also been used to inter structures. Though shicra are found at some locations outside the North Central Coast, such as El Paraiso and Buena Vista in the Chillon Valley, they consistently date to the Late Archaic (Benfer 2012; Haas and Creamer 2012; Quilter 1991). For archaeologists, shicra are not only a diagnostic construction technique but also an ideal tool for radiocarbon dating (Ruiz et al. 2007; Shady et al. 2001).

Shicra

In most reports about Late Archaic sites, shicra are not mentioned, or they merit a few lines indicating their presence or absence in a particular

Figure 7.3. Shicra holding construction fill at Caral.
Photo courtesy of Håkan Svensson.

building or construction level. Most archaeologists who mention shicra approach them from a purely functional standpoint—they were simply the chosen method to move the stone fill from the quarry or riverbed to the construction site (Shady et al. 2001). One of early analyses proposed that the technique afforded extra stability and protection from earthquakes in this seismically active region (Huapaya 1977).

When there has been an effort of interpretation for shicra, it has consistently been framed in the ongoing debates about social complexity that dominate the discourse for this region. Feldman (1980, 1985) has argued that the monumental construction at Aspero is the evidence of a chiefdom-level society and proposes that the shicra were used to measure contributions toward a labor tax owed to the ruling elite. Quilter (1991) offers the alternate suggestion that the society was mainly organized around a kinship system and that shicra may have been used by family or kin groups to complete construction or renovation on discrete areas for which they were responsible. The use of shicra in the ritual interment of ceremonial rooms has been linked to the existence of an elite class with power over ceremonial systems (Benfer 2012). These arguments tell us much about archaeologists' preoccupations with social and political organization but very little about the shicra themselves or the lives of the people who made and used them.

I argue that a closer look at the variability of shicra makes it clear that they were more than simply a way to move rocks from point A to point B.

While the use of shicra is highly diagnostic of Late Archaic architecture, they were not used in every monumental structure at every site or during every construction phase in any given building's history. At Bandurria, Alejandro Chu's (2008) excavations of Monticulo 1, which has the typical terraced-mound shape, central staircase, and associated sunken circular court of other Late Archaic sites, did not uncover any shicra. At Aspero, shicra are used throughout the construction levels of Huaca Alta and Huaca de los Idolos, but they only appear in the most recent construction of Huaca de los Sacrificios (Feldman 1980:49). Ruiz et al. (2007) report that in the Proyecto Arqueólogico Norte Chico's investigation of eight sites with monumental architecture in the Pativilca Valley, all but two of the monumental structures were built using shicra.

Not only does the appearance of shicra vary within and between sites, but the character and use of the shicra also vary. At Pampa San Jose in the Pativilca Valley, Ruiz et al. (2007) report finding shicra filled with sand and soil. Shicra filled with sand, soil, or fine gravel appear at other sites. These shicra must have been filled in situ because the openings of the looped netting are too large to hold fine materials for carrying. At Caballete in the Fortaleza Valley, Creamer et al. (2013:24) report a shicra containing small stones that had a lining of leaves on the inside that prevented the stones from falling through the weave. At El Paraiso, Quilter (1991) reports an unusual large stone accompanied by offerings discovered inside a wall in Unit 1. One of the offerings was a miniature shicra filled with cakes of lime wrapped in leaves (Quilter 1991:423–424). Alejandro Chu (2008) discusses a unique find at Bandurria where shicra were not found in the monumental architecture, but instead a "proto-shicra" was found in a domestic context. This proto-shicra was found in the fill of a platform in a domestic area. Chu (2008) notes that this shicra was more simply constructed than most shicra found in monumental contexts and that it was made of junco, a local reed that is less durable than the usual *S. totora*. Shicra also vary widely in size, from shicra at Vichama that were so large that they could not have been carried when filled with rocks (William H. Isbell, personal communication 2012) to the miniature shicra reported by Quilter. Other sites, such as Aspero, have shicra that were placed empty into construction fill.

Furthermore, while shicra are common, highly visible in architectural profiles, and clearly diagnostic of Late Archaic architecture, they are in fact not very well documented. Most articles and reports will note either the presence or absence of shicra with little further elaboration. Even unusual cases, such as those mentioned above, rarely earn more than a few sentences in publication. Thus, the variability of shicra is likely underestimated and certainly poorly documented. Yet with the limited published data that are available, it seems clear that shicra, at least at some times and in some places, served more than purely practical purposes.

Another important point to note is that the practical value of shicra is not in fact obvious. Shicra are most frequently used in renovation where they are placed either inside a room, as a base for new construction, or between an existing and a newly constructed wall. This means that the shicra are not necessary to hold the rock fill in place; the existing masonry can do this job quite well. In fact, in many cases shicra are intermixed with loose fill in a single depositional event. But if shicra are not structurally necessary, and if their use is so variable, we must ask why shicra are used in some places and not others.

Shicra as Equipment and the Praxis of Shicra

In the case of shicra, we see a ubiquitous type of material culture that has been practically absent from archaeological analyses. Archaeological interest in the Late Archaic sites of the North Central Coast has been overwhelmingly focused on ideational accounts of the monumental architecture. The relatively abrupt appearance of large scale construction has been used primarily to advance various theories about the level of social complexity during the Late Archaic period in the Central Andes; the architecture itself has been treated as indicator of and proof for different configurations of social and political hierarchy. By remaining tightly focused on a limited portion of the archaeological record, these interpretations run the risk of falling into the category of analysis that reveals more about the preoccupations of the archaeologist and their political and intellectual concerns than the concerns and lives of the ancient people (Brück 2005; Quilter and Koons 2012). By focusing exclusively on the exercise of power by a proposed elite class, they foreclose on other possible social systems and ignore the lived experience of people who are not a part of the proposed elite. An important part of this lived experience is the origin and development of the monumental architecture over time. By addressing only the final state of the monuments as they are seen today, archaeologists have neglected the dynamic nature of the monumental construction and the ongoing experience of their development on the landscape.

What might be gained by broadening our analysis and attending to the unremarkable parts of the material culture? If we focus on the material, rather than the ideal, if we apply the lessons of Heidegger and of ecological psychology, and if we ask "how" rather than "why," what new understandings might emerge?

Shicra can be understood as equipment, or tools, in Heidegger's sense of the word. That is, for archaeologists, as well as the ancient people of the North Central Coast, shicra were/are things of concern. As such, shicra exist within a totality of equipment and are referential to the other equipment in this totality. In addition, for the ancient people this equipment primarily would have been perceived in an embodied manner through interactions with the body. For archaeologists, it is also appropriate and necessary to approach the investigation of shicra in this manner, as Heidegger specifies that phenomenological investigation arises out of praxis. So, the understanding of shicra properly starts with the physical experience of shicra and the totality of equipment to which they belong.

Shicra are made from S. *totora*, which grow in the river valleys of the North Central Coast and were an indispensable industrial resource for the people of the Late Archaic. These strong sedges were used not only to make shicra but also baskets and bags, as well as mats that were used for a variety of purposes, including roofs, walls, and burial of the dead (Chu 2008; Feldman 1980; Quilter 1991; Shady 2004; Willey and Corbett 1954). Through the material connection of S. *totora*, shicra reference all of these as part of the totality of equipment.

In addition to the materials, the technique for making the shicra is referential to another vitally important part of the Late Archaic material culture—cotton nets. A variety of weaving and twining techniques were used to make cotton nets of various sizes. Nets were indispensable for catching fish, an essential component of the Late Archaic diet. Nets are common in the archaeological record and show a great deal of skill and care in their construction and diversity in details, such as overall design, size of the openings, knotting techniques, and repair techniques (Feldman 1980, 1992; Moseley 1975; Quilter 1992; Willey and Corbett 1954). Fishing nets are not only important for people living close to the ocean, at sites like Aspero and Bandurria, but also for inland residents. Excavations at Caral, for example, have revealed that marine resources were an important component of the diet and must have been brought inland from the coast on a regular basis (Shady 2004; Shady and Caceda 2008).

We should also think about the experience of making the shicra, which is part of the embodied engagement with the shicra. Shicra would have necessarily been made by at least two people working together, since the looped net design requires the shicra to be filled as it is made. The experience of making shicra, then, was necessarily a cooperative experience, one that required coordination and agreement to create a final product that met the needs of the project. We can imagine many questions, such as who might have worked together; did the team have only the one job of making the shicra; or did they have other responsibilities? What were the possible attitudes of the people who may have happily engaged in a family project excitedly working for the favor of a deity or reluctantly labored under the watchful gaze of some task master? In fact, we will never know how it felt, in this sense, to experience making the shicra. We can, however, know that the material requirements of the task demanded at least two people to work closely together in a cooperative project; this was not an experience of solitude but an experience that required each member of the team to rely on the other in a coordinated rhythm of work.

With this very short consideration of the experience of the shicra, we already see that the context of the shicra automatically references items (equipment) and actions (praxis) that were of deep concern to the ancient people, and we know how these actions and items were linked in distinct and concrete ways. We may not know exactly how work was divided or organized, but we do know that people harvested S. *totora* and prepared it for the manufacturing process in a way that would have been physically the same whether the end product was shicra, baskets, or mats. In this activity, they took advantage of the specific material properties of S. *totora*. Through the interaction of the material and the skilled actor, affordances emerged: the renewable nature of the resource that allowed it to be harvested year after year and the strength and flexibility of the plants that allowed them to be woven, looped, braided, and shaped to meet specific needs. Preparing S. *totora* was an experience that could end with shicra for a massive construction project, a basket to hold food for your family, large mats to help shelter community members from the beating summer sun or the endless damp winter wind, or a smaller mat to wrap a deceased child and a few offerings of shell in preparation for burial. The S. *totora*, in knowing hands flexed and bent, expanded and contracted to meet many needs in the community.

The skill needed to make the looped netting was related to similar techniques used to create other bags and nets of junco, S. *totora*, or cotton. The skill in constructing each was to make the right decisions in material, in specific technique, and especially in size to capture, hold, and contain the items needed whether rocks, fish, or some other item. We do not know if every person had these skills, or how they were passed from generation to generation. But we do know that daily life, especially for people living on the coast, was filled with physical references to the work of catching and containing through the physical presence of nets in the community and in the actions of people engaged in the work of making nets, repairing nets, and using nets. These references were also present in the bodies of these people whose hands and arms, backs and shoulders, and eyes were physically shaped by and through realizing the work of catching and containing. This concern for and relationship with nets extended into the past before the first monuments appeared in the North Central Coast, as maritime resources, including fish that are captured in nets, were important to the diet of coastal peoples from about 9200 BC (Stackelbeck 2011:200).

The Questions We Ask and the Work They Do

How did shicra form part of a dynamic system that included human bodies and behavior? With this very short exploration of shicra as part of a totality of equipment and the bodily experience of shicra, we can clearly see that shicra also do the physical work of catching and containing. Whether the shicra are filled with rocks, or sand, or tiny cakes of lime, the shicra are doing the same work. With this understanding, it is not necessary to ask why use shicra when they are not necessary from an engineering perspective. This is simply another way of asking what function do the shicra serve. By reference to their contexts we know what the shicra do, how they work, and what function they serve. If we excavate a shicra that does not catch or contain any object apparent to the archaeologist, this need not change our understanding of how the shicra work. Instead, we must acknowledge that whatever the shicra contain is not a part of the archaeological record, either because it was not preserved or because it was never a part of the material record.

Further, we can understand that the shicra are made of a material that built a broad network of references to a wide variety of items that were used daily. And, this network of actions involved people, as well as the daily labor of people. In fact, the totality of equipment to which shicra referred includes the most mundane and basic items, the items that were made, used, repaired, and discarded as a matter of course with one exception, and that exception is the monument. Here we see the way in which an approach informed by phenomenology and the ideas of ecological psychology repositions the archaeological analysis. Rather than asking why people began to build these structures, we have instead started with the question of how these structures fit into people's lives. Rather than an ideal approach that asks what ideas drive this construction, we have started with materiality and praxis. *How did people experience this construction?*

The answers to these questions are radically different for two reasons. First, only one set of questions is answerable given the data and techniques currently available. As archaeologists working in this specific context, without written records, with very little representational art, and with no appropriate ethnographic models, we cannot answer the questions of "why" and "what idea." So we attempt to fit the data into theoretical models that provide answers on a multiple-choice format; if there is monumental architecture, it is either A—a chiefdom—or B—a state. These models reflect, necessarily, our modern experiences, concerns, and contexts. But, the second set of questions, questions of praxis and context, can be answered with the data and techniques available, even if only in an imperfect way. And the answer will not be multiple choice; nor will it, if we are honest and careful with our phenomenological methods, only reflect our experiences and concerns. It can instead reflect the context and experiences of the Late Archaic people, what Ingold might call a taskscape.

So what do shicra tell archaeologists about the monuments of the North Central Coast? They tell us we must understand them as part of a daily lived context. The monument is a new thing, but it is created in a completely and entirely comprehensible manner that has fidelity to the past and the present for its creators. People create the monument by capturing and containing with totally ordinary materials to create equipment that performs a daily type of work. This equipment and the work it does is completely familiar in this context because it has already been part of the lives of these people for generations.

The shicra that are part of the monument do the work of capturing and containing, but because the shicra are inside the filled-in rooms and between the walls, they themselves are captured and contained. The difference between the shicra and the building, the constituent part and the whole, can be thought of not as a difference in function but rather a difference in material properties. In fact the monument is a product of the affordances of rock, the solid persistence and rigid nature of stone, as compared to the flexible and malleable S. totora. In the totality of its context, the shicra, the rock, and the skilled bodies of the people, we can see that the monuments themselves physically performed the same work as the shicra, the same work as the people who made and used the cotton fishing nets and S. totora baskets, only on a different scale. What is captured is different from before, and the persistence of what is captured may have been a novel experience. But there is no reason to believe that the construction of monuments necessarily changed the lives of the people of the North Central Coast in any radical way. There is no need for recourse to an explanation for some extreme change in social or political systems, especially as the monumental architecture itself is the only evidence for such proposed transformations.

What has been seen as a disruption, as an indicator of crucial change in the social and political field, can be understood, within its context, as part of a dynamic system and as an organic development within that system. The continuity of the shicra within its context is the indicator of the continuity of the lived experience in the community. The archaeological record bears out this interpretation, as we see the continued use of shicra in monuments of very similar architectural style for more than a thousand years, often at the same site. Much like the fishing nets, generations of people lived with, created, renewed, and repaired monumental constructions in a daily way, year after year.

It is tempting to think of the shicra and the objects they contain as bundles in Pauketat's (2013) sense of a thing that binds its sensuous qualities to other things in a form of entanglement. In fact, the physical action of the shicra, the monuments, and the people involved in their construction, which I have described as capturing and containing, aligns very closely with his idea. The impulse is particularly strong in the case of the unique shicra occasionally reported, whether extremely large or extremely small, and containing not construction fill but dirt, or lime, or nothing at all. In fact, perhaps there was a symbolic or ritual component to the shicra, and in creating the shicra people were creating bundles that incorporated memory, or ceremony, or some other unknown type of power. But if we are to remain true to the terms of the theory and method laid out for this study, we must leave the symbolic aside and think instead about how the material properties of these things, the embodied experiences of their creation, and their use allowed them to evoke the past, shape the present, and frame the future.

It is clear that we lack a number of tools and experiences that would make this investigation complete. A program of experimental archaeology would perhaps give more insight into the phenomenological experience of making the shicra, but there are countless possibilities for how to organize this work that could leave many options for the timing, duration, and rhythm of the task. Furthermore, the bodily engagement with shicra is different for the ancient and the contemporary person; we are doing different things with the material. Finally, we must acknowledge that, even if we are not attempting an ideal interpretation, the archaeologists' lack of social and cultural knowledge will impact their analyses even at the level of praxis. This is a limitation of phenomenology rarely discussed by its advocates, and it is seen as a fatal flaw by its critics. Yet even with this limitation, this brief case study has shown how phenomenology can be a useful way to approach certain kinds of questions in archaeology.

Conclusion

In their discussion of using the phenomenological method in the cognitive sciences, Gallagher and Zahavi (2012) highlight the importance of suspending the *natural attitude* as a first step in phenomenological methodologies. This means that the phenomenologist must step away from the conviction that an objective, utterly knowable world exists for the understanding. This is the basic assumption underlying our Western scientific thought and our research designs and methods in archaeology.

Adopting a method like the one described here, which attempts to transcend the subject-object divide and engage with the way in which people and objects are in the world together, allows us to suspend the natural attitude and launch a different type of analysis than what has generally been produced in archaeology. The advantages are that we can approach our investigations from a fresh perspective and ask new questions that reflect the entire context of the past, not just a piece of the past that has been carved into narrowly defined, abstract categories. However, this will be a very different type of analysis than we are used to seeing. Many will find such attempts unsatisfying.

In the current case study, the focus on shicra allows us to see the emergence of monumental architecture in the Late Archaic period on the North Central Coast of Peru in a way that attends to daily life, connects the emergent monumental architecture to traditions of the past, and recognizes the role of people and the objects they create in forming a history. But for the scholars who have been engaged in the ongoing debates that have tried to characterize the social and political structures of this period, this case study may seem tangential. It does not stake a claim, or affix a label—"this is a chiefdom" or "this is a state." But I contend that its conclusions are no less important, and that if we use this case study to rethink the question of social complexity, we can see new possibilities. The emergence of monumental architecture has been used as a sign and a proof of increased social complexity and emergent political power. This has been because, more than anything, it is a new thing in this landscape. But as the current case study shows, the monumental architecture may not have been experienced as something totally new; through the material nature of the monumental architecture it had strong connections with people's daily lives and with traditions of work and material culture from the past. Further, the aspects of the monumental architecture that are most unique, namely its scale and permanence, can be seen as emerging out of the affordances of the materials, the *S. totora*, and the stone in interaction with the unique embodied skills of the people. When the question is approached from this angle, a new possibility opens up; it may not require a new political power to make a monument. It may simply grow out of the way that people and things are together in the world.

Acknowledgments

This chapter developed from my Master's thesis work and would not have been possible without the guidance, support, and wisdom of William H. Isbell and Ruth Van Dyke. The insight and enthusiasm of my fellow contributing authors allowed me to develop a vague and muddled idea into what I hope is a thought-provoking contribution, and for this I am deeply grateful. Many thanks go to Anne Hull for drafting the map of the North Coast. I must also express my special appreciation to Dr. Ruth Shady Solís and the Proyecto Zona Arqueológica Caral for the opportunity to work at the Aspero site. Of course, any omissions or errors in this text are my own.

References Cited

Aho, Kevin. 2009. *Heidegger's Neglect of the Body*. State University of New York Press, Albany, New York.

Barrett, John, and Ilhong Ko. 2009. A Phenomenology of Landscape: A Crisis in British Landscape Archaeology? *Journal of Social Archaeology* 9:275–294.

Barrett, Louise. 2011. *Beyond the Brain*. Princeton University Press, Princeton.

Benfer, Robert A., Jr. 2012. Monumental Architecture Arising from an Early Astronomical-Religious Complex in Peru, 2200–1750 BC. In *Early New World Monumentality*, edited by Richard L. Burger and Robert M. Rosenswig, pp. 313–363. University Press of Florida, Gainesville.

Boivin, Nicole. 2008. *Material Cultures, Material Minds*. University of Cambridge Press, Cambridge.

Brück, Joanna. 2005. Experiencing the Past? The Development of a Phenomenological Archaeology in British Prehistory. *Archaeological Dialogues* 12(1):45–72.

Chemero, Anthony. 2009. *Radical Embodied Cognitive Science*. The MIT Press, Cambridge.

Chu Barrera, Alejandro. 2008. *Bandurria: Arena, mar y humedad en el surgimiento de la Civilización Andina.* Proyecto Arqueólogico Bandurria—Huacho. Huacho, Peru.

Creamer, Winifred, Álvaro Ruiz Rubio, Manuel F. Perales Munguia, and Jonathan Haas. 2013. The Fortaleza Valley, Peru: Archaeological Investigation of Late Archaic Sites (3000–1800 BC). *Fieldiana Anthropology* 44:1–108.

Cummings, Vicki, and Alasdair Whittle. 2004. *Places of Special Virtue: Megaliths in the Neolithic Landscapes of Wales.* Oxbow Books, Oxford.

Feldman, Robert. 1980. Aspero Peru: Architecture, Subsistence Economy, and Other Artifacts of a Preceramic Maritime Chiefdom. Unpublished PhD dissertation, Department of Anthropology, Harvard University, Cambridge.

———. 1985. Preceramic Corporate Architecture: Evidence for the Development of Non-Egalitarian Social Systems in Peru. In *Early Ceremonial Architecture in the Andes,* edited by Christopher B. Donnan, pp. 71–92. Dumbarton Oaks, Washington, DC.

———. 1992. Preceramic Architectural and Subsistence Traditions. *Andean Past* 3:67–86.

Fleming, Andrew. 1999. Phenomenology and the Megaliths of Wales. A Dreaming Too Far? *Oxford Journal of Archaeology* 18:119–125.

———. 2006. Post-Processual Landscape Archaeology: A Critique. *Cambridge Archaeological Journal* 16(3):267–280.

Gallagher, Shaun, and Dan Zahavi. 2012. *The Phenomenological Mind,* 2nd ed. Routledge, New York.

Gell, Alfred. 1998. *Art and Agency: An Anthropological Theory.* Clarendon Press, Oxford.

Gibson, James J. 1979. *The Ecological Approach to Visual Perception.* Erlbaum, Mahwah, New Jersey.

Haas, Jonathan, and Winifred Creamer. 2006. Crucible of Andean Civilization: The Peruvian Coast from 3000 to 1800 BC. *Current Anthropology* 47(5):745–775.

———. 2012. Why Do People Build Monuments? Late Archaic Platform Mounds in the Norte Chico. In *Early New World Monumentality,* edited by Richard L. Burger and Robert M. Rosenswig, pp. 289–312. University Press of Florida, Gainesville, Florida.

Haas, Jonathan, Winifred Creamer, and Álvaro Ruiz. 2004. Dating the Late Archaic Occupation of the Norte Chico region in Peru. *Nature* 432(23/30):1020–1023.

Heidegger, Martin. 1982 [1954]. *The Basic Problems of Phenomenology,* Translated by Albert Hofstadter. Indiana University Press, Bloomington, Indiana.

Huapaya Manco, Cirilio. 1977. Vegetales como elementos antisísmico en estructuras prehispánicas. *Arqueología PUC* 19(12):27–38.

Hurcombe, Linda. 2007. A Sense of Materials and Sensory Perception in Concepts of Materiality. *World Archaeology* 39(4):532–545.

Ingold, Timothy. 2000. *The Perception of the Environment.* Routledge, London.

———. 2005. Comments on Christopher Tilley: *The Materiality of Stone: Explorations in Landscape Phenomenology. Norwegian Archaeological Review* 38(2):122–129.

———. 2007. Materials Against Materiality. *Archaeological Dialogues* 14(1):1–38.

Keane, Webb. 2005. Signs Are Not the Garb of Meaning: On the Social Analysis of Material Things. In *Materiality,* edited by Daniel Miller, pp. 182–205. Duke University Press, Durham, North Carolina.

Küchler, Susanne. 2005. Materiality and Cognition: The Changing Face of Things. In *Materiality*, edited by Daniel Miller, pp 206–230. Duke University Press, Durham, North Carolina.

Latour, Bruno. 2005. *Reassembling the Social: An Introduction to Actor-Network Theory*. Oxford University Press, Oxford.

Macann, Christopher. 1993. *Four Phenomenological Philosophers*. Routledge, New York.

Michaels, Claire, and Claudia Carello. 1981. *Direct Perception*. Prentice-Hall, Englewood Cliffs, New Jersey.

Miller, Daniel. 2005. Materiality: An Introduction. In *Materiality*, edited by Daniel Miller, pp. 1–50. Duke University Press, Durham.

Moseley, Michael. 1975. *The Maritime Foundations of Andean Civilization*. Cummings Publishing Company, Menlo Park, California.

Olsen, Bjørnar. 2010. *In Defense of Things: Archaeology and the Ontology of Objects*. Altamira, Lanham, Maryland.

Pauketat, Timothy R. 2013. Bundles of/in/as/time. In *Big Histories, Human Lives*, edited by John Robb and Timothy R. Pauketat, pp. 35–56. School of Advanced Research Press, Santa Fe, New Mexico.

Pinney, Christopher. 2005. Things Happen: Or, From Which Moment Does That Object Come? In *Materiality*, edited by Daniel Miller, pp. 256–272. Duke University Press, Durham, North Carolina.

Pozorski, Sheila, and Thomas Pozorski. 2008. Early Cultural Complexity on the Coast of Peru. In *The Handbook of South American Archaeology*, edited by Helaine Silverman and William H. Isbell, pp. 607–631. Springer, New York.

Pozorski, Thomas, and Sheila Pozorski. 2012. Preceramic and Initial Period Monumentality within the Casma Valley of Peru. In *Early New World Monumentality*, edited by Richard L. Burger and Robert M. Rosenswig, pp. 364–398. University Press of Florida, Gainesville.

Quilter, Jeffrey. 1991. Problems with the Late Preceramic of Peru. *American Anthropologist* 93(2):450–454.

———. 1992. To Fish in the Afternoon: Beyond Subsistence Economies in the Study of Early Andean Civilization. *Andean Past* 3:111–125.

Quilter, Jeffrey, and Michele Koons. 2012. The Fall of the Moche: Critique of Claims of South America's First State. *Latin American Antiquity* 23(2):127–143.

Ruiz Rubio, Álvaro, Winifred Creamer, and Jonathan Haas. 2007. *Investigaciones arqueológicas en los sitios del arcaico tardío (3000 a 1800 anos A.C.) del valle de Pativilca, Perú*. Instituto Cultural del Norte Chico, Barranca, Peru.

Shady Solís, Ruth. 2004. *Caral, the City of the Sacred Fire*. Centura SAB, Lima.

Shady Solís, Ruth, and Daniel Caceda Guillen. 2008. *Áspero, la cuidad pesquera de la civilización Caral: Resuperamos su historia para vincularla con nuestro presente*. Proyecto Especial Arqueologio Caral-Supe, Lima.

Shady Solís, Ruth, Jonathan Haas, and Winifred Creamer. 2001. Dating Caral, a Preceramic Site in the Supe Valley in the Central Coast of Peru. *Science* 292(5517):723–726.

Smith, Adam T. 2003. *The Political Landscape: Constellations of Authority in Early Complex Polities*. University of California Press, Berkeley and Los Angeles.

Smith, Linda B. 2005. Cognition as a Dynamic System: Principles from Embodiment. *Developmental Review* 25:278–298.

Stackelbeck, Kary. 2011. Faunal Remains. In *From Foraging to Farming in the Andes*, edited by Tom Dillehay, pp.193–204. Cambridge University Press, Cambridge.

Thomas, Julian. 1996. *Time, Culture and Identity*. Routledge, London.

Tilley, Christopher. 1994. *A Phenomenology of Landscape*. Berg, London.

———. 2004. *The Materiality of Stone: Explorations in Landscape Phenomenology*. Oxford University Press, London.

Vega-Centeno, Rafael. 2005. Consumo y ritual en la construcción de espacios públicos para el eriodo arcaico tardío: El Caso de Cerro Lampay. *Boletín de Arqueología PUCP* 9:91–121.

———. 2010. Cerro Lampay: Architectural Design and Human Interaction in the North Central Coast of Peru. *Latin American Antiquity* 21(2):115–145.

Willey, Gordon, and John Corbett. 1954. *Early Ancon and Early Supe Culture*. Columbia University Press, New York.

CHAPTER EIGHT

From Banned Bodies to Political Subjects

Immigrants in Protest Bundles

Jessica Santos López

In 2006, massive protests for immigration reform took place across the United States. Five years later, in the comparably large-scale Occupy Wall Street (OWS) Movement, banners displayed "Full Citizen Rights for all Immigrants" in many variations. The presence of immigrants as protagonists, coprotagonists, or supporters in public protests is evidence of a shift toward greater diversity in the composition of human and workers' rights movements in the United States. These protests pose a challenge to the public, to government authorities, and to the participants themselves regarding the rights usually attributed to citizenship. This chapter analyzes the materialization of choreographed masses of bodies that challenge the current power structure in public spaces and destabilize notions of the legal. The relational, physical, and material properties of the bodies of legal and illegal subjects is in conversation with several key areas of current theory, including phenomenology and the politics of space, bundled aspects of human and object engagement, and the paradoxical political and legal presence/absence of undocumented immigrants. From a phenomenological perspective, "oft-repeated embodied experiences and material practices . . . continuously constitute and reconstitute cultures and identities. . . . Landscapes, objects, and beings are components of experiential fields simultaneously socially constructed and constructive. Daily practices and performances, as acts of 'inhabitation' or 'dwelling,' construct landscapes of experience" (Pauketat 2013:35).

Phenomenology renders possible the understanding of the dialectical relationship between the body and the world (Merleau-Ponty 1965 [1945]). Perception is possible through bodily senses that are experienced by a subject in time and space, whose body is itself an expressive space. When undocumented immigrants participate in protests, unequal power relations inform the relationships among bodies and objects in space and influence perceptions, transactions, contentions, and negotiations. These power relations become evident in the way different and differentiated bodies move in relation to other bodies, in the places assigned to them, and in the places (un)lawful bodies claim in the public sphere.

Subjects give meaning to their practices and experiences and the places where they occur. For the undocumented immigrant, these everyday transactions are enmeshed and shaped by the conditions of their existence. Immigrants' actions and interactions are informed by the consciousness of their legal status—their rights of citizenship. Giorgio Agamben (1995) developed the concept of *bare life* to describe a human being (a body) deprived of rights. Agamben's *homo sacer* refers to those who are banned or reduced to bare life by the apparatus of power that differentiates between good and bad citizens. Homo sacer is a paradoxical figure who stands outside, as well as within the law; to be excluded, homo sacer must be legally defined. In the modern state, those with bare life, as well as those with citizenship, are subject to political control. The distinction between citizen and homo sacer poses fundamental questions regarding the modern state's power to produce lawful personhood. The state and law, by practicing the power to separate political beings from bare life, has given the undocumented person the status of sacer. "The exclusion of the sacred man from the community, the absence of bare life from the political realm . . . becomes today . . . the unthinkable limit—which is also the very foundation—of the dysfunctional society in which we live" (Kishik 2012:77).

From an objective perspective, a police officer and a protester could be considered equivalent, but the immigrant becomes political and not simply ontological through the exercise of violence. The undocumented immigrant is transformed from a banned body into a political subject when asserting his or her rights in a space assigned to the citizen. Independently of legal status and statutes, the distinction between homo sacer and citizen is highly porous when it comes to performing politics. Despite sovereign power-control strategies, the undocumented can employ collective tactics to reconfigure the common sense that informs their *being-in-the-world* by challenging the state's power of regulating life in society. Thus, there is analytical potential for the understanding of bodies and objects in protests as *bundles.*

The concept of bundles is a useful construct to further the understanding of the simultaneity and interplay of objects and subjects present in protests in public spaces. Bundles enable the analysis of "the processes of creation of complex objects and the interactive roles of their components . . . [and their potential] for expanding the understanding of connections among objects and societal concepts and practices" (Zedeño 2008:363). Things, places, and bodies are closely associated or bundled in meaningful ways within a material dimensionality (Keane 2005). Thus, protests could be considered bundles where the relationship between individual bodies, objects, and public places is fundamental in constituting and reconstituting identities and the ways they are perceived. These processes also lead toward a transformation of the subject-object divide because they are mutually constitutive. On the one hand, bodies can be constrained by the limitations of space. On the other hand, I consider the body a subject-object whose material qualities are essential to the desired political impact. "The very physicality of the body imposes a schema on space through which it may be experienced and understood. An experience of space is grounded in the body itself; its capacities and potentialities for movement" (Tilley 1994:16).

Using photos and videos uploaded to the web by immigrant participants and supporters in the 2006 Immigration Reform Protests and the 2011 OWS Movement, I discuss the ways in which the materiality of bodies intersects with the materiality of spaces during public protests. I argue that the participation of undocumented immigrants in public protests is a result of a long series of common experiences, discursive practices, collaborative and conflictive exchanges, and even concurrent actions. For this reason, I first focus on a protest where undocumented immigrants and their supporters were the central figures. Then, I look at the immigrants' participation in the movements with a more diverse composition and a wider structural social critique. In both cases, the undocumented immigrants' participation in a public collective demonstration could be characterized as a result of a series of liminal states, directly involved in the negotiation of subject-object relationships and activating future liminal moments. This state of liminality is also illustrated in the space itself, and it is two-fold. On the one hand, the perception of public space is transformed according to the subjects' decisions to interact with a particular material world and the way that interaction is represented to a greater audience. On the other hand, it is illustrated in physical and governmental constraints that each space imposes on the body. In theory, the body of the citizen is allowed to interact freely with the public material world, while the body of the noncitizen is at worst a trespasser and at best an excess. Nonetheless, these bodies are constantly interacting with the public and private material world in complex ways, whether as domestic workers in private homes or as political actors in semipublic spaces. Through an intentional selection of the material world as a public sphere for the performance of politics, and the indistinguishable copresence of noncitizen and citizen bodies, these *protest bundles* become a challenge to the space-control mechanisms of the state. As illegal/banned political subjects engage with the material world, they redefine the political.

To elucidate these processes, this chapter consists of three parts. The first part discusses the dialectical relationship between the material and subjective qualities of the body through a phenomenological approach in which the concept of bundles is elaborated in terms of its analytical potential for understanding bodies with human-thing qualities and of human-object relationships in *protests as bundles*. The second part discusses the performance of situated collective bodies where contributions from urban anthropologists will be extended to analyze the politics of public space. Finally, I analyze the possibilities of modifying structures of power relations through collective actions where undocumented immigrants become political subjects by embracing the power of their physicality to affect their material world.

The Body: Its Subjectivity and Materiality

From Plato to Descartes, philosophers describe the body as the imperfect holder or prison for the mind. By contrast, phenomenologists transcend this dichotomy, as they describe how everyday life demonstrates the relational interconnection between mind and body. The body is subject (a site of perception), as well as object (in direct material relation with the outside world). A phenomenological approach creates space to understand the body of the undocumented immigrant as one that intentionally relates to the world but that also shares time and space with other objects. A human tends to understand her/his "being in terms of that being to which it is essentially continually, most closely related — 'the world'" (Heidegger 1962 [1927]:14). According to Heidegger, *Dasein* — literally translated as being-there, being-in-time — is constituted by an everydayness embedded in temporality. Heidegger proposes the concept of dwelling "to link place, praxis, cosmology, and nurture" (Tilley 1994:13). Undocumented bodies and subjects intentionally position, perceive, and place themselves in the world, but they do so in response and in relation to states' control mechanisms. However, the intentionality of those actions occurs as part of a situated experience. The places where these processes occur are known to these subjects and consequently are imbued with meaning, whether symbolical or practical. Simultaneously, these spaces of interaction are a product of the interests of different actors, public and private, who have gradually implemented a series of control mechanisms. Public and semi-public spaces, understood as the public sphere where politics take place, are not meant for undocumented immigrants.

Protests as Bundles

Bodies—documented and undocumented—are the objects of power-control mechanisms and are enmeshed in power relations. State-imposed illegitimacy informs objective and subjective aspects of daily practices and identities as parent, sibling, worker, spouse, and so forth. Nonetheless, the human-thing relationships that are part of performing these roles contain social and material dynamics beyond the law, and all these different roles constitute a bundle in itself. The undocumented immigrant navigates a terrain of resistance/negotiation that entangles her or him in a constant process of *becoming* within a series of relational fields. These embodied and relational practices unfolding in time make the undocumented immigrant simultaneously inside and outside the normative arrangement of citizenship. As Yael Navaro-Yashin points out (2009:9), the "horizontal and two-dimensional imaginary of 'the network,' has to be complemented with a theory of sovereignty and history, which introduces qualified verticality and multiple dimensionality."

In a protest where undocumented immigrants are a main component, there are complex interactions between objects and subjects and quasi objects and quasi subjects in a network where humans, discourses, and objects are interwoven (Latour 2005). Undocumented bodies as subjects and objects interact with other "legal" and "illegal," semilegal and semi-illegal bodies and political subjects, as well as the rest of the *props* present in the *mis-en-scéne* (theatrical setting) of a public protest. These bundles, which are assembled or entangled to reinforce the intended message, include oppositional forces, absent supporters, and personal and spatial biographies. Hence, power struggles are constitutive of public demonstrations and must be accounted for when analyzing protest bundles. Relational qualities are by no means static or consistent, because "every act, motion, practice, or experience bundles something" (Pauketat 2013:39).

Each of the relational objects that are bundled in a protest possesses different properties, power, and use value. Each fulfills different purposes in time and space. Some subject-objects are intrinsic to the protest bundle, such as bodies, props, law enforcement officers, and public spaces. Some become important at different stages of the protest. The presence of others, such as media, art forms, or riot squads depends on the circumstances.

These subject-object categories can create more subcategories and different sets of interactions with particular communicative properties. Each protest has its own unique qualities depending on time, space, country, subject-object composition, and frequency of interaction. Bundles are a theoretical construct based on transcending subject-object and mind-body dichotomies. The material dimensions of practice, performance, and experience transcend the momentary to create bundles of biographical and meaningful associations with historical implications. People, places, and things mediate in different ways the fields or networks of relationships in which they exist, and each exert a different influence.

The material bodies of undocumented immigrants are intrinsic to the fabric of most of the cities and urban areas in the world. These noncitizens struggle with other political subjects for the state and law to recognize their presence as subjects that deserve rights and responsibilities equal to any "good" and "well-behaved" citizen as legally defined. Political acts occur within the material world and transform the perception of space, the human subjects that are part of it, and the observers while problematizing legal categories of personhood.

Collective Bodies in Public Space

Bourdieu (1977) refers to the systematic exclusion of some groups from spaces of social power as "symbolic violence." Once political subjects, documented and undocumented, have intentionally challenged exclusionary practices through a protest, how does their performance of collective bodies transform the *field*, a structured social space? According to Bourdieu, the *habitus* regulates, reproduces, and produces practices. A consequence of this regulation is a common understanding of the world reinforced by an equally common understanding of everyday practices (Bourdieu 1977:79). The concept of habitus also adjusts to objective potentialities in a given situation and serves to elucidate the possibilities for modifying power relations through collective actions. In the case of immigrants' rights, participants in immigrant protests produce practices through dynamic interactions in the social field in which this symbolic violence takes place and consequently transform the habitus and the identity of the agents, albeit slowly and subtly. Protest bundles are one of the most conspicuous ways through which people struggle to transform the habitus. They challenge understandings of the composition of any field by transforming perceptions of participants, as well as observers.

Urban anthropologists' analyses of the relationships among subjects, objects, and public space provide significant insights into political participation in the public sphere. Neil Smith and Setha Low (2006:3) define public space as "the range of social locations offered by the street, the park, the media, the Internet, the shopping mall, the United Nations, national governments, and local neighborhoods." They argue that neoliberalism, on the rise since the 1980s and the events of 9/11, have redefined public space in the United States to resemble the privatized commons of conservative seventeenth- and eighteenth-century countries. Accordingly, in the contemporary United States, "property owners and consumers in the marketplace are the new citizens" (Smith and Low 2006:2). By contrast, David Harvey (2006:17) refers to the "'public sphere' as an arena of political deliberation and participation, and therefore as fundamental to democratic governance." All three authors consider the imagery of the Athenian agora as a powerful reference for the practice of democratic ideals, but which in practice was, and still is, an exception instead of the rule.

Smith and Low's aim is to underscore the repoliticization of public spaces while calling attention to the dichotomy of the public-private space. They argue "the lost geography of the public sphere comes with a concurrent loss of politics, however partial. Abstracting from the location of real events and social relations removes an entire dimension of political relationality" (Smith and Low 2006:7). With a similar emphasis on the potential of urban public spaces, Harvey (2006) points out that the city changes according to political circumstances, and the way it is collectively understood and lived transforms accordingly. To this effect, he stresses the study of transformative and relational processes between the multiplicity of spaces, things, and subjects and their mutual and heterogeneous constitution (Castree and Gregory 2006:157).

I contend that the analysis of undocumented immigrants as noncitizen political bodies contributes to a better understanding of space and the relationship between objects and subjects. A relevant example is the 2003 Immigrant Workers Freedom Ride; eighteen chartered buses from across the United States convened in Washington, DC, and then delivered their occupants to join 125,000 protesters in New York City. De Genova (2009:450) argues that "the migrant activists of the Freedom Riders defiantly asserted their subjectivity, first of all and most significantly, with their bodies." This civil rights activist group negotiated the interstitial space between bare life and political life. The undocumented immigrants seized their democratic rights to free speech by occupying visible public space, as in the 2006 Immigration Reform and 2011 Occupy Wall Street protests. Through their strategy of becoming moving targets, these migrants produced a "differential space" whereby they reformulated the political through embodied dissidence. The state no longer had—if it ever had—complete control over public space. According to De Genova (2009:450), "the itinerant assemblage of the Freedom Riders' dissident and deportable bodies did not merely 'express' the protest of unruly subjects who repudiated their own objectification by the state's regime of migrant securitization. It *enacted* and *performed* a crucial dimension of subjectivity itself—its material and practical corporeality." These protests push "the question of the speaking subject front and centre [and] provoke fundamental questions about politics [and] . . . who can be political" (De Genova 2009:451).

Becoming Political Subjects: Immigrant Protests

I use the stylistic device of the vignette to describe two public demonstrations, or protest bundles, where undocumented immigrants participated in various ways. The first vignette is from the 2006 Immigration Reform Protests, which were a response to the then-proposed Border Protection, Anti-terrorism, and Illegal Immigration Control Act of 2005 (H.R. 4437) that would criminalize anyone aiding undocumented immigrants and substantially reinforce immigration control mechanisms. Here, the participation of undocumented immigrants and their supporters was fundamental to the protest itself. The second is from the 2011 OWS Movement where on Wall Street in New York City protests were organized demanding fair distribution of wealth and targeting the wrongdoings of corporations and financial institutions aided by the U.S. government. Here, undocumented immigrants and their supporters were incorporated as part of a broader coalition that involved immigrants participating as individuals and as members of collectives, such as the Immigrant Worker Justice Working Group. Both vignettes portray protests as products of spatial and temporal bundles of human-object relationships.

I developed these vignettes using online images that appear when one searches the Internet for "2006 Immigration Reform Protests," "Occupy Wall Street New York," "Occupy Wall Street Immigrant," and "Occupy Zuccotti Park." After a visual account of these protests, I provide a series of discursive statements, written and verbal, from participants, passersby, activists, and scholars and then an analysis. I include these statements to help contextualize the events while allowing a more complex understanding of the forces in play within both movements. My focus is on pro-immigrants' rights protesters within OWS, because OWS's diverse composition represents well the radical potential of protests as bundles. Thus, the main objective is to initiate an understanding of the forces and processes in play in terms of the participation of undocumented immigrants in protest bundles.

Vignette I: 2006 Reform Protests

Millions of people participated in these protests in the major cities of many states. The largest demonstrations occurred in Chicago, Los Angeles, and Dallas where the numbers of participants ranged from 100,000 to 500,000 people. The protests lasted for three months. The streets were full of bodies in motion, music performances, megaphones, and thousands of colorful banners and flags. From the outside, the protests resembled a choreographed mass moving in the same direction. That mass engulfed buildings, streets, and sidewalks. There was no space for any means of transportation. There were banners in all colors, fonts, and sizes in Spanish and English:

> We are America.
>
> We Didn't Cross the Borders, The Borders Crossed Us.
>
> We Work Hard, We Pay Taxes.
>
> Immigration is Not a Crime.
>
> Stop Racism.
>
> We Work for America.
>
> No Mas Deportaciones.
>
> Dignity has No Borders.

The 2006 protests exhibited and restated two main issues among undocumented immigrants: first, a split identity (illustrated in the presence of countless U.S. flags), and second, the right to work.

Vignette II: 2011 Occupy Wall Street

Most readers are probably familiar with televised and Internet images of the OWS protest in Zuccotti Park, a publicly accessible private park in Lower Manhattan. These images generally depict a community-like, festive atmosphere. According to *The New Yorker*, "it is a dinner party of sorts, albeit one with donated, often organic food served on paper plates. There's tea, too, of course, mostly herbal—rooibos and chamomile. . . . At first glance, Zuccotti looks to a casual visitor like a crowded, messy homeless encampment. But it doesn't take long to discern an earnest, underlying orderliness. The plaza is loosely divided into 'centers,' each watched over by members of its own 'working group'" (Hertzberg 2011).

But when one searches for images depicting immigrant participation in OWS, the images convey a different tone; they are more militant than the 2006 protests for immigration reform. They reveal aggressive exchanges between the police and protesters, during which many young people were arrested, battered, or tear gassed by the police.

In Spanish and English, banners proclaimed the following messages:

We are the 99%.

Due to recent budget cuts the light at the end of the tunnel has been turned off.

Immigrants Occupy!

Wall St. is War St.

One Day the Poor will Have Nothing to Eat but the Rich.

Stop Deportation Now.

Where's My Bailout?

Workers' Rights are Human Rights.

Immigrant Rights are Human Rights.

Rights for the People, Not Corporations.

War Targets Poor People of Color.

Workers Say NO to Racist Police Murders.

Full Rights for All Immigrants.

A broader set of demands reflected a more diverse composition of protesters, although here I have highlighted some of the immigration-related banners. Also, because of the length of the New York City protests and the places where they occurred, there was greater diversity in the ways the demonstrations unfolded. Some were more violent than others; the attendance varied, as well as the festive environment; and some protests were theme specific, such as the one on International Migrants Day on December 18, 2011.

The OWS protests were not strictly immigrant rights demonstrations. However, the protestors' critiques were directed toward the corporate personae of banks, the financialization of daily practices, precarious working conditions, and the interrelated maintenance of racial hierarchies (Selkirk 2013); these concerns affect citizens and noncitizens alike (Graeber 2013). The rise of the corporate prison-industrial complex has become a major focus of criticism for immigrant rights groups. Within both sets of protests, undocumented immigrants articulated a claim to be treated with dignity as human beings and workers. They deftly connected their daily predicaments to a broader and deeper social solidarity challenging the power structure that defines them as the banned *Other.*

Protest Bundles

Bodies intersected with objects and spaces—constituting and modifying each other—during undocumented immigrant protests (Figure 8.1). Although the OWS Movement was more heterogeneous than the 2006 Immigration Reform Protests, the trope of the protest bundle can help illuminate both cases.

A sampling of the range of subject-objects that constituted OWS protest bundles included the following items: banners of all sizes; mic checks; assemblies and working groups; police officers in cars or on horses; police dogs; blogs; (online) interviews; theater, music, and other performances; Zuccotti Park; Bryant Park; trees and flowers; tents; journalists with microphones, cameras, and cell phones; mattresses; information and food tables; bikes (locked or on the ground); vendetta masks; families; people of all ages, cultures, and subcultures; cigarettes; drinks; pizzas delivered; books; and Hula-Hoops and other toys (Table 8.1).

Each of these categories and subcategories could be understood as a bundle with its own relational dynamics. By way of illustrating the flexible and transient relationships in play, think of an undocumented person in a protest with a banner and chance attention from the media. The protester

Figure 8.1. OWS first anniversary performance in Zuccotti Park. Photo by Mona Kareem.

has agency over what is written on the banner, but the banner may give a different message to an audience if framed differently by the media. An undocumented immigrant surrounded by a multitude could feel safer walking, speaking publicly, or expressing herself/himself through a banner. However, the media exists to inform its audience about struggles, such as protests; or, more cynically, the corporate media apparatus exists to sell events. Moreover, ideologically partisan media coverage may spin the immigrant as a friend or as a criminal, as part of a community, or an invading Other. Depending on the ideology of the audience, the message may be received critically or not. The protest will be supported or condemned, and the images of participants will be used to profile and prosecute persons (documented or not).

These diverse compositions of subjects and objects have to be understood as part of the politics of public space. According to Smith and Low (2006), city squares have witnessed political struggles where their appropriation by private capital has undermined democratic practices. For example, the material and symbolic space of Union Square in New York

Table 8.1. Object-subjects found in OWS protest bundles

Object Category	Subcategories
Bodies	Citizens, non-citizens, and those with intermediate legal status; individuals of all ages, ethnic backgrounds, sizes, heights, and shapes; the bodies of law enforcers, families, and members of subcultures.
Props	Banners, flags, flyers, informational material, bikes, tables, and food.
Art Forms	Music, musical instruments, theater, performance art, and games (on street and stage).
Media	Cameras, microphones, mobile phones, television, and radio.
State Control Mechanisms	Law enforcement officers in any type of transportation (cars, horses, or bicycles), detectives, undercover police, riot squads, other specialized units, arms and ammunition, and chemical weapons, such as tear gas.
Places	Wall Street, Union Square, Zuccotti Park, and Bryant Park.

City has been used to display the Union Army during the Civil War, to protest the Vietnam War, and to protect homeless rights. After 9/11, many congregated there to collectively outpour their feelings.

> [In] a genuine and spontaneous, if rare, expression of the public sphere that for a while operated outside official control. After barely two weeks, the New York Police Department began to reimpose state authority by invoking specific regulations, beginning at the margins—geographically, legally, and politically—and working their way to the center until this unprecedented expression of the public sphere, occupying and remaking public space, was eventually closed down. (Smith and Low 2006:12)

The two-month-long occupation of Zuccotti Park followed this tradition of materially and symbolically occupying public places . . . except for the fact that Zuccotti Park is not a public space. It is owned by Brookfield Properties, although it was set up as a public plaza by the owners. Thus, the end of the occupation and the eviction of the protestors necessitated an orchestrated effort between Brookfield Properties and the state. Mayor Bloomberg pronounced that "unfortunately, the park was becoming a place where people came not to protest, but rather to break laws, and in some cases to harm others" (NBCNEWS.COM, November 16, 2011).

In an analysis of OWS, Silvia Federici (2011) contends that "the very concept of 'occupation' is connected with the tactics that students [in response to the commercialization of education] adopted over the last two years, from New York to Berkeley and beyond, and especially in Europe." She also associated the "tent cities" created by evicted people in the United States with the Bonus Army of the 1930s and the Poor People Campaign in the 1960s (Federici 2012). For her, all these instances are the production of a collective imagination.

There were many other spaces throughout the city that were used, if not occupied, by different groups coordinating the movement. Because undocumented immigrants run the risk of being deported if arrested, the locations and frequencies of regular meetings added another layer of concern. Undocumented persons must make strategic decisions about how to engage with places and material surroundings as they pursue their protest goals.

Discussion

The 2006 and 2011 protestors stood together against the hierarchical power relations that organize things and subjects, and they challenged their own erasure and negation by corporate forces and the state (Figure 8.2). Especially in the 2006 Immigration Reform Protests, the immigrants dwelling in every corner of every city, big and small, were suddenly rendered visible. It became impossible to ignore their presence. They became "undocumented, unafraid, and unapologetic." The undocumented immigrants who lived and worked quietly in cities, while consuming and enjoying them, were now demanding to be recognized as part of society and its productive economy. In the last decade, we have witnessed how undocumented immigrants have gone from practicing a silent/hidden agency to a physical/spoken agency with the support of many individuals and sectors of society and in collaboration with other minority and oppressed groups.

The diverse composition of the protesters in both events transcends the legal distinction between the documented and the undocumented. My summary of the situated and embodied participation of undocumented immigrants is that "we have the right to a dignified existence despite and because of our sameness and Otherness to you, us, and ourselves." This is exemplified by one of the concerns coming from immigrant groups in OWS through an online interview with *The Real News Network*. Thanu Yakupitiyage, from OWS Immigrant Worker Justice Group and the New

Figure 8.2. T-shirt on OWS first anniversary participant.
Photo by Mona Kareem.

York Immigration Coalition, explains her support of OWS while expressing her discomfort with the marginal role of peoples of color in the movement (Yakupitiyage 2012). In her critique, she highlights the predominance of white, middle-class people at the beginning of the movement. As a result of discussions about the underrepresentation of people of color, organizers changed the "Declaration of the Occupation of New York City" (Declaration) to incorporate statements about racial justice. The Declaration, which denounced corporate forces, is an example of the product of resistances and negotiations among subgroups holding the same general objectives. Federici (2012) argues that immigrants' rights were a substantial part of OWS. Like Thanu Yakupitiyage, she differentiated between two phases in the movement as the movement sought to overcome gendered and racial divisions. However, Federici perceives a change from more to less diversity. At the beginning, OWS had "a substantial presence of women and people of color both in the working groups and in the decision-making processes" (Federici 2012). Subsequently, two interrelated factors made the crowd more homogenous: first, the police attacks, and second, the spread of the movement to neighborhoods. The media, political activists, organizers, and educators helped resignify the field to the protester and the observer.

Conclusion

Protest bundles are collective, public performances involving the bodies of documented, as well as undocumented Others. The undocumented took this opportunity to move their bodies from private into public spaces made relatively safe through anonymity. Through participation in protest bundles, undocumented immigrants challenged the power structures that keep them in the shadows vulnerable to all kinds of violence. By moving their bodies in public spaces, they claimed the freedom *to be* to the eyes of everyone, but also to themselves. Undocumented immigrants experienced the strength of being together, bundled, as part of a multitude where "culture articulates conflicts and alternately legitimizes, displaces, or controls the superior force. It develops in an atmosphere of tensions, and often of violence, for which it provides symbolic balances, contracts of compatibility and compromises, all more or less temporary" (de Certeau 1984:xvii).

For undocumented immigrants, the decision to participate in a public protest is the result of a long series of experiences, practices, and discursive exchanges. Undocumented immigrants have gone through many transformations throughout the time that they have been thinking, living, and acting in a specific context. Their habitus is confirmed by the interaction of subjects in a particular material world where they have been subjected to what they have come to realize are unacceptable living and working conditions. But, habitus is susceptible to change, and it is changed, however subtly. The history of these changes and exchanges are embodied in the undocumented immigrants bodies, daily practices, daily struggles and tactics, and in their participation in broader political movements where they are co-creators of new experiences and new forms of *being* in and to the world.

Undocumented immigrants have intentionally established a subversive language nourished by popular culture, democratic practices, and networks that connect objects and subjects—the local with the global and vice versa. Moreover, with the OWS Movement, they claimed that they were as subordinated as any other citizen, and they were political beings. Protesters joined forces to claim the rights and benefits of an ideal of citizenship, of belonging, instead of the definition provided by law. Even in public spaces informed by mechanisms of state control, protesters resignified and transformed those city squares with their corporeal presence. No one can view Zuccotti Park the way they did before OWS, nor will they be able to forget the lessons of collaborative exchanges and social power. The strategies employed by the state to control the population could not overcome the tactics used by "dominated" subjects in public spaces.

Subversive language and practices are developed in the shadows, where I argue it is possible to create and nourish a space of radical anonymity with greater consequences. The apparent absence of the undocumented immigrant in the eyes of the law can be better characterized as an anonymous presence where she/he subverts a power structure by being copresent with others who hold a legal status. On the one hand, the state and law acknowledges undocumented immigrants, albeit in a negative form, in the various mechanisms of population control arranged and deployed by the state. On the other hand, the imperceptible presence of these political subjects renders them perceptible to those that are witnesses through different means, physically and virtually.

The citizen and the noncitizen are bundled through an embodied and spatial experience. The police officer, the protester, and the documented and undocumented constitute a bundle while being part of other bundles, too. Within a protest bundle, their sometimes contentious, tense calm, or even momentary collaborative coexistence transform each other and the way they experience and understand the world. Cohabitation makes these transformations possible. But, cohabitation was possible in the first place because of the protesters rights and claims to perform politics in public space and because of the government's dispositions and internal negotiations to uphold freedom of speech while maintaining public order. This common dwelling—particularly in sustained long-term protests—is a result of human negotiations that push the limits of law. For example, OWS protesters argued that they were protected by the First Amendment to the U.S. Constitution, and for that reason there was no need to comply with the permit requirements for public protests stipulated by law. This is an example of how "subjectivity and objectivity connect in a dialectic producing a *place* for *being* in which the topography and physiography of the land and thought remain distinct but play into each other as an 'intelligible landscape'" (Tilley 1994:14).

I argue for a type of research that prioritizes the understanding of "large-scale and long-term ways that history plays out" (Pauketat 2013:37) through bundling experiences. In terms of space, undocumented bodies have claimed a place in private and public spheres. The latter was evidenced in the 2006 Immigration Reform Protests and then in the OWS Movement. I suggest, along Kishik's analysis over the Israeli Palestinians "present-absent" status, that undocumented immigrants,

> those present-absent people—because of (rather than despite) the deep ambivalence that permeates their everyday lives, and without forgetting their suffering—could be conceived as the true vanguard of a community still to come. What we are dealing with here is far more than a bizarre legal status. We may expand the present-absent condition to encompass anyone who, by choice or by necessity, partakes *and* abstains, is included *and* excluded, is there *and* not there, when a certain place, rule, institution, association, occupation, culture, language, or practice is at stake. (Kishik 2012:79, emphasis in original)

Finally, I argue that bundles enable those who live and work in the margins to elude control mechanisms of the state, to "effectively govern . . . [their] bodies, control . . . [their] actions and manipulate . . . [their] desires" (Kishik 2012:86). Immigrants are undeniably and irredeemably bundled, aiming for human, not merely legal, status.

Acknowledgments

First and foremost, I would like thank Ruth Van Dyke for her enthusiasm, encouragement, and vision. This project would not have been possible without her support from its initial (playful) stages until the final touches while keeping an unwaning academic rigor. I also thank my colleagues in this edited volume who earnestly engaged in each step of this journey. I would like to take this opportunity to thank the Clifford D. Clark Diversity Fellowship for Graduate Students at Binghamton University – SUNY. Finally, I dedicate this publication to Dr. Carlos Buitrago Ortiz (1930—2013), my mentor, colleague, and Distinguished Professor at the Anthropology Department, University of Puerto Rico. Buitrago inspired me (and countless others) to pursue a career in academia and become an anthropologist. I will be ever grateful to him for teaching me the meaning of dedication and passion for learning, always accompanied by "Ay, comay!" Finally, I could not address the economic and political aims of the Occupy Movement as an outcome of broader economic and political processes. However, this phenomenological approach is an initial stage of a larger project, which I undertake as part of a wider activist network.

References Cited

Agamben, Giorgio. 1995. *Homo Sacer: Sovereign Power and Bare Life*. Stanford University Press, Palo Alto, California.

Bourdieu, Pierre. 1977. *Outline of a Theory of Practice*. Cambridge University Press, Cambridge.

Castree, Noel, and Derek Gregory. 2006. *David Harvey: A Critical Reader*. Blackwell Publishers, Oxford and Malden, Massachusetts.

de Certeau, Michel. 1984. *The Practice of Everyday Life*. University of California Press, Berkeley and Los Angeles.

De Genova, Nicholas. 2009. Conflicts of Mobility, and the Mobility of Conflict: Rightlessness, Presence, Subjectivity, Freedom. *Subjectivity* 29:445–466.

Federici, Sylvia. 2011. Feminism, Finance and the Future of #Occupy. An Interview with Silvia Federici by Max Haiven. *InterActivist Info Exchange*, on November 26, 2011.

——. 2012. We Are Witnessing the End of an Era. Interview with Silvia Federici by Max Hanninger. *InterActivist Info Exchange*, on October 11, 2012.

Graeber, David. 2013. *The Democracy Project*. Spiegel and Grau, New York.

Harvey, David. 1973. *Social Justice and the City*. Johns Hopkins University Press, Baltimore.

——. 2006. The Political Economy of Public Space. In *The Politics of Public Space*, edited by Setha Low and Neil Smith, pp. 17–34. Routledge, New York.

——. 2012. Ahead of May Day, David Harvey on Urban Uprisings. *Democracy Now*, April 30, 2012.

Heidegger, Martin. 1962 [1927]. *Being and Time*. Translated by J. Macquarrie and E. Robinson. Blackwell, Oxford.

Hertzberg, Hendrik. 2011. A Walk in the Park. *The New Yorker*, October 17. New York.

Keane, Webb. 2005. Signs are Not the Garb of Meaning: On the Social Analysis of Material Things. In *Materiality*, edited by D. Miller, pp. 182–205. Duke University Press, Durham.

Kishik, David. 2012. *The Power of Life: Agamben and the Coming Politics*. Stanford University Press, Palo Alto, California.

Latour, Bruno. 2005. *Reassembling the Social: An Introduction to Actor-Network-Theory*. Oxford University Press.

Merleau-Ponty, Maurice. 1965 [1945]. *Phenomenology of Perception*. Translated from French by Colin Smith. Routledge and Kegan Paul, New York.

Navaro-Yashin, Yael. 2009. Affective Spaces, Melancholic Objects: Ruination and the Production of Anthropological Knowledge. *Journal of the Royal Anthropological Institute* (NS) 15:1–18.

Pauketat, Timothy R. 2013. Bundles of/in/as Time. In *Big Histories, Human Lives*, edited by John Robb and Timothy R. Pauketat, pp. 35–56. School for Advanced Research Press, Santa Fe.

Ryans, Natasha. 2011. Protesters In, Tents Out at NYC "Occupy" Park. *NBC*, November 16. New York.

Selkirk, Alexander. 2013. My Encounter with the "Precarious and Service Workers Assembly" of Occupy Oakland. *InterActivist Info Exchange*, January 30, 2013.

Smith, Neil, and Setha Low. 2006. Introduction: The Imperative of Public Space. In *The Politics of Public Space*, edited by Setha Low and Neil Smith, pp. 1–16. Routledge, New York.

Tilley, Christopher. 1994. *A Phenomenology of Landscape: Places, Paths and Monuments*. BERG, Oxford, UK.

Yakupitiyage, Thanu. 2012. Occupy Wall St. Takes Up Immigration Reform. *The Real News Network*. YouTube Interview. Jan 12, 2012.

Zedeño, María Nieves. 2008. Bundled Worlds: The Roles and Interactions of Complex Objects from the North American Plains. *Journal of Archaeological Method and Theory* 15:362–378.

CHAPTER NINE

Materiality as Problem Space

Mark W. Hauser

Of all the changes of language a traveler in distant lands must face, none equals that which awaits him in the city of Hypatia, because the change regards not words, but things.
  —Italo Calvino, *Invisible Cities*

Italo Calvino's *Invisible Cities* is part travelogue and part thought experiment. It imagines conversations between Kublai Khan and Marco Polo, and the reader is asked to experience fifty-five fantastical avatars of Venice where the laws of physics do not dictate how people inhabit them. In Hypatia, a mercurial instantiation, Polo learns that "signs form a language, but not the one you think you know" (Calvino 2006 [1978]:43). Polo's discussion of Hypatia highlights the limitations of language and meaning in understanding experience and bodily practice. Things wholly constitute the city's life. They have a superficial congruence with bodily experiences but at the same time beguile its interpretation. As such, Hypatia emphasizes two questions central to *Practicing Materiality*. In what ways do things, and the way we talk about them as scholars, make the people? We can also flip this question around. How do we populate strange places through things, language, and their combinatory potential?

These two questions are not particularly novel. However, they do help clarify how past and present actors operate through matter. In some ways I see this volume, to which I am adding a brief and cursory comment, as a response to Daniel Miller's (2005) edited volume *Materiality* and as a reflection on the 2013 meetings of the American Anthropological Association (AAA) in Chicago. Miller's volume assembled case studies, objects, ideas, and stances to demonstrate "why we need to engage with the issue of materiality as far more than a footnote or esoteric extra to the study of anthropology. The stance to materiality also remains a driving force behind humanity's attempt to transform the world in order to make it accord with beliefs as to how the world should be" (Miller 2005:2). The concept of materiality had been, for more than a decade, an often-used but poorly defined analytic in material culture studies, anthropology, and cultural geography (see Hicks [2010] for an excellent historiography of archaeology's engagement with material culture studies, materiality, and the material turn). It resonated with embodiment, practice, experience, and context, yet for all of the carefully crafted studies, few had attempted to systemically address the heart of materiality: its immateriality. Contributors to this volume have a different goal: to put materiality into practice and in so doing make materiality material.

After all, it is a material world, and most of us don't quite know what that means. This was made absolutely clear during the 2013 AAA in Chicago. Some panels explicitly tackled relational ontologies and the kinds of distributed agencies that they imply. Some participants, like Bruno Latour, described and critiqued assumptions about object-human boundaries and the mechanism through which people are acted upon in the material world. They described the potential dangers of uncritically applying ideas from late modernity backward. They also described the promise of an approach in which the mapping of relationships in and about the world—without direction, without privilege, and without category—provides an empirical base through which to map its contours. Others called for an anthropocentric view highlighting the need to maintain a critical apparatus that focuses not only on the forces that shape human history but also the experiences of the forces at more intimate scales. The question of a relational ontology's novelty was discussed. It could be claimed that it approaches the world dialectically, just without archaeology's critical apparatus.

Perhaps one of the reasons why there is such difficulty in mapping the units, categories, and scales of analysis in materiality is that at the core there is a potential false dichotomy. There is, of course, a dialectical approach—materiality with the development of thesis, antithesis, and synthesis (McGuire 2013; Miller 2005). While not novel, there is always value in mapping how objects can be alienated and alienating. There is also a phenomenological approach to materiality—an approach followed by many of the authors in this volume. Practice, embodiment, and indeterminacies focus analytical attention on structuring and structured matter. Perhaps it may be too optimistic to suggest this, but such approaches are not mutually exclusive. Addressing the materiality of slavery requires an absolute attention to its political economy. Yet to overlook the embodied experience and how that experience structured everyday life of the enslaved would also miss that story. Rather materiality's promise is to focus attention on political economy at the human scale.

By asking the question, "so, just how do I operationalize [materiality] in my work?" (Van Dyke, this volume), authors confront a central problematic to archaeological practice: *How do we make the immaterial material?* For authors of this volume materiality is best operationalized through phenomenological and anthropocentric approaches—specifically the use of one's own sensory experiences to constitute a context in which past actors were limited, as well as enabled. They ask us to imagine the kinds of coordinated efforts to raise *shicras* (dry stone terraces) during the Late Archaic in North Central Peru or the repeated gestures in saint shrines located in southern Turkey where the shrine and the person being venerated are indistinguishable. In the American Southwest, a pot screams as it breaks against the floor, and blue and white faience condition moralities in the eighteenth-century Portuguese Empire. Tasks recounted through these material interactions are made particularly difficult for a number of reasons: the plant-fiber nets containing the shicras cannot be fabricated in advance but must be woven as the dry stones are put in place; Saint veneration in Southern Turkey carries with it political risk in a landscape dominated by Sunni Islam and Turkish nationalism; the places where materials for making pots are located are charged with symbolic and spiritual dangers; and the Portuguese Empire contained highly diverse populations with multiple interests at play. Through their engagement, the contributors to this volume demonstrate the rich benefits of the employment of theory and empirical detail across a broad canvas.

I enter this commentary as a scholar who investigates the contributions of Africans and their descendants living under conditions of slavery. Sidney Mintz and Richard Price famously observed that the central contradiction of race-based slavery was that "slaves were legally defined as property; but being human they were called upon to act in sentient, articulate, and human ways" (Mintz and Price 1992:25). Mintz and Price were referring to people as things. Things and people were more than just intertwined, they mutually constituted the institution of slavery. As such I have approached the material turn with a degree of ambivalence. On the one hand archaeologists working on sites associated with enslaved labor in the West Indies are confronted on a daily basis with the power of objects and things. Slavery as practiced in early seventeenth-century Spanish Jamaica and Hispaniola, as harsh as it was, looked only vaguely like the chattel slavery practiced in late eighteenth-century British Jamaica and French Hispaniola (Higman 2005). Early colonial settlements, such as Sevilla Nueva, highlight a certain ambiguity in the classifications of others and their commodification through law (Woodward 2011). Yet the genome of later plantation societies based on asymmetrical social relations were present in the sixteenth century. Despite its many permutations, through a process of philosophical extension, racial thinking, and enlightenment reasoning, slavery moved from a process of manipulating canon and royal law to account for the ownership of human beings (Las Casas) to a process where the definition of humanity was manipulated in order to make Africans property (Graeber 2006).

Historians, anthropologists, and archaeologists working on sites associated with eighteenth-century enslaved labor in the West Indies are confronted on a daily basis with the power of objects and nonhumans under conditions of slavery (Figure 9.1). "Things" evoked pasts, transformed lives (for better or worse), and obscured realities. People built fortunes through the acquisition and transformation of matter into things. People were also subject to those fortunes as humans were commodified through market activities. Working in the cane fields was dangerous and energy consuming.

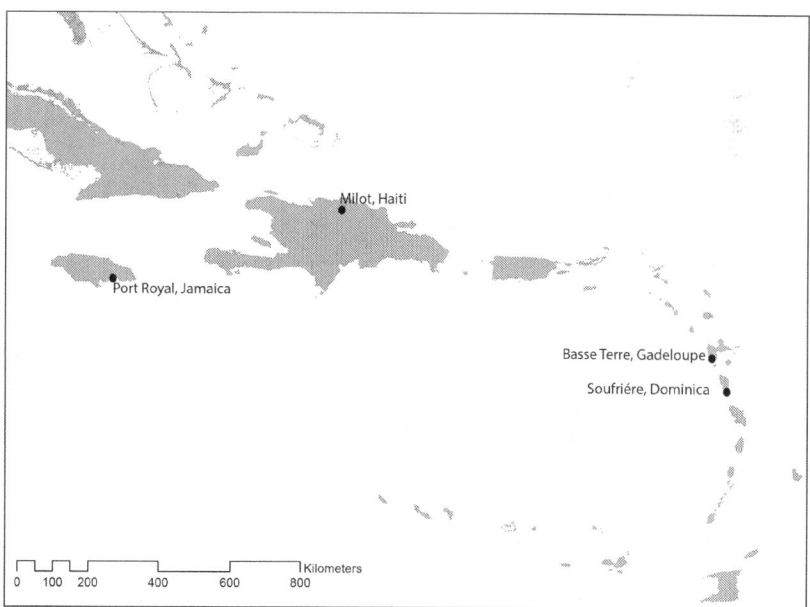

Figure 9.1. Locations of sites discussed. Drafted by Mark Hauser.

The ability to capture and store potable water meant the difference between life and death, especially in the Caribbean, which is plagued by both hurricanes and drought (Hauser 2014). The pottery jars and glass vessels in which laborers stored their water broke frequently, especially as people transported water to and from the field (Arcangeli 2015). Broken sherds become a testimony to the lives of these laborers and the adaptive strategies they deployed to make a new life under harsh conditions (Armstrong 1990; Handler and Lange 1978; Wilkie 2000). They speak not so much to slavery but its effects embodied in the material practice of everyday life. My small contribution to this field was to interpret the presence of these objects socially (Hauser 2011). Specifically, I stressed the centrality of their distribution and the way its scale reflected a large island-based economy largely organized by enslaved laborers and higglers. They were ubiquitous in slave contexts; but they were also hidden in plain sight. Such texts through which we construct past lifeworlds can be obscured or overshadowed by more readily accessible information that might reinforce dominant narratives that reference the absence of material contributions rather than their presence.

In short, slavery had a material record, and that record can be found in the archaeological landscapes of the Americas. The materiality of slavery, however, is fraught with absences born out of asymmetrical social relations. As such, I frame these comments around what is for me a concern: How does one practice materiality in light of slavery? How does one take into account the uneven nature of archaeological evidence? How do we privilege a human-centric point of view without imposing our agendas informed from contemporary stances? And how do we truly engage in such systems at a human scale? Here, I review some of the main themes that emerged in the volume: power, agency, and experience. In my conclusion I talk about materiality as a *problem space* and point to some continuing trends and perhaps my own questions.

Power

The first theme that resonates in all the chapters in this volume is that any consideration of objects with agency is incomplete without a direct engagement with power relationships. Materiality insists that humans exist in a world partially constituted of and by objects. For some, (Gosden 2005; Gosden and Marshall 1999; Joyce 2003; Mills and Walker 2008) examining how objects accumulate biographies (see Kopytoff [1986] for life histories) over the course of their lives has provided a venue to describe the situated interactions between objects and humans and the meaningful ruptures and congruencies these encounters create. Still others have contributed to materiality by describing configurations of time and space in objects and material practices that surround them (Hicks 2010) in a manner reminiscent of Bakhtin's (1981) *chronotope*. Here I focus on one aspect of the turn that has gained some purchase in historical archaeology and anthropology—symmetrical archaeology—to show how the essays in this volume challenge some of its basic assumptions. Reaching back to Bruno Latour's, John Law's, and Michael Callon's wholesale reshaping of the project of sociology (Callon et al. 1986; Latour 1993, 2005; Law and Hassard 1999), symmetrical archaeologists have insisted on the absolute equity of humans and things (Olsen 2010; Webmoor and Witmore 2008:59; Witmore 2007). By bringing symmetry to humans and nonhumans, they hope to evade dualisms implicit in Western philosophical traditions of the past three hundred years: subjects and objects; humans and nonhumans; people and things (Boivin 2008; Hodder 2012; Olsen 2010; Shanks 2007). This last trend is the one with the greatest implications, especially when one considers the unintended repercussions of the theoretical registers we adopt.

Power is something left out of, or minimized in, many such studies. Take for example Ian Hodder's most recent monograph, *Entangled*, on the relationship between humans and things. This book nicely maps at larger scales the degree to which humans and nonhumans become increasingly imbricated in densely packed networks. In western Asia, approximately 8,000 years ago, humans were tied up in entanglements that reached beyond the human scale but were still immediately perceivable for the subject involved. Fast forward 7,800 years when humans were caught up in a much more complicated and distantly configured network in which sugar provided one cluster. Sugar provided quick energy for factory workers in England and demanded large amounts of labor to be found in Africans and their descendants in the West Indies (Mintz 1985). Sugar was entangled with other commodities, such as coffee, tea, and the vessels used to drink them. The consumption of these goods brought about sumptuary rituals, which created difference and assembled relations of power along commodity chains.

Here, entanglements become interesting ways to map particular geographies. They offer a departure from meaning, intentionality, and causality. They are not however, necessarily sufficient for understanding the workings of change. In his careful discussion of commodity networks in the eighteenth-century Atlantic, Hodder fails to mention power—the primary category of analysis in Mintz's (1985) *Sweetness and Power*. The book was about a larger narco-capitalism, but it was also about the simultaneous disciplining of labor on both sides of the Atlantic through the production and consumption of sugar. This is a potential pitfall in most approaches that privilege the relationship between human and nonhuman vs. human and human. This approach can be critiqued in much the same manner as Liz Brumfiel critiqued systems analysis of environmental anthropology in the 1970s and 1980s. As Brumfiel (1992) argued in her critique of this approach, gender, class, and faction are more than just preoccupations of the modern, they meaningfully situate human experience in a larger political economy of ideas and things.

For her part, Erina Gruner shows how ritual paraphernalia "act" in the form of bundled objects to create solidarities over vast distances, instantiate a powerful theocratic elite in the American Southwest, and distribute knowledge within and between kin groups. Her scope is ambitious but aided by the deep archaeological literature in this area. Corporately owned ritual paraphernalia contain objects with their own animacy and other objects that accumulate qualities from their makers. These paraphernalia transmit knowledge through care and maintenance. They also confer elite status on the specialists who most effectively activate them. For Gruner, the comparison of two burials, the "Magician" from Ridge Ruin and Burial 14 from Pueblo Bonito, create an intriguing puzzle when thinking about bundles. The two burials located in different culture areas from different times contain similar ritual paraphernalia. Add to that the individuals with whom the paraphernalia are associated shared a similar life history suggesting "replication of objects, but also the replication of the social role of the warrior, the same role held by ethnohistoric Pueblo stick swallowers." While this could lead to often-criticized tropes of stable identities, Gruner focuses on the destruction, distribution, and reassembling of paraphernalia from the tenth century to the present when such knowledge and the status conferred by its position disappears (Gruner, this volume). In so doing, she is able to fashion a social history that centers the object and the accumulated experiences surrounding it.

According to Rui Gomes Coelho, ceramics shaped, disciplined, and socialized imperial subjects as they circulated through the networks and territories associated with the Portuguese in Iberia, Africa, Brazil, Japan, and the Estado de India. Networks are interesting ways to map particular geographies, but as Coelho notes they are not necessarily sufficient for understanding the workings of governance. Juridical and representational practices bound administrators, commissioned artisans, and owners together. These networks varied in intensity and density and can reveal certain choke points. In a political organization as widely spread and diverse as the Portuguese empire of the seventeenth and eighteenth centuries, this diversity raises a question about what is significant to the various *actants*. Coelho asks to what extent were the moral obligations associated with the circulation, manufacture, and use of clay body material venues for encoding imperial moralities onto colonial subjects? This morality is essential to the formation of the modern state and the inequalities on which it is based.

Slavery also suggests potential blind spots in key analyses of consumption like Norbert Elias's (1994) *The Civilizing Process*. While the fork Elias discusses is a well-known example of how courtly practices shaped taste through networks of power and prestige, the colony was never seriously considered in his analysis. Take a mahogany chair (Figure 9.2)—when did it become the fashion of the courtly elite to employ it in their drawing rooms? Was it before or after they were in the drawing rooms of planting elites in Martinique, Vietnam, India, or Guyana?

Figure 9.2. Mahogany canework bergère chair used by Napoleon Bonaparte (1769–1821) now at Maidstone Museum, Kent. Courtesy of WikiCommons under Creative Commons Attribution 3.0; photographer, Linda Spashett.

Agency

A second theme in this volume is a defense of anthropocentricism in the face of recent attempts to abandon it (see Olsen 2010; Webmoor and Witmore 2008; Witmore 2007). That is to say, the contributors to this volume insist on a privileging of human agency in human systems. Attempting to develop categories of analysis that are most relevant to the past lives we are trying to describe in a bit more detail is central here. As each author acknowledges in this volume that there is a particular trap in anthropocentrism in which the relationship between humans and nonhumans can reify a Western ideal/real ontology. There is a productive tension, however, in marinating an anthropocentrism while at the same time exploring the contours of the human and nonhuman relationship. Take the not-so-simple concept *agency*. Authors practicing materiality in this volume have been largely guided by Alfred Gell's concept of *secondary agency*.

In *Art and Agency*, Gell (1988) locates in art objects systems of action that transform the world rather than encode symbolic propositions about it. Objects are embedded in a network of social relations. Agency is mediated by indexes, which are matter that motivates responses, inferences, or interpretations. Indexes stand in variegated relations with prototypes, which inform their creation, artists who fabricate those objects, and recipients who activate them. As such it offers a departure from meaning, intentionality, and causality. According to Webb Keane, objects develop a kind of agency in their combinatory potential. For Keane, "icons and indexes in themselves, 'assert nothing'" (Keane 2003:418–9); they are ordered by a further sign. Bundles bring together the icons, indexes, and most importantly the assumptions that make metaphors work. Such bundles can be found in ritual objects, but they can also be located in the mundane. Bundles, however, are not timeless entities. Their composition changes, their use changes, and their meaning changes or sometimes gets lost. A phenomenological approach, through which the relationship between humans and things can be re/created, situates the human scale of materiality, while the bundles they inhabit extend our scales of analysis and illuminate the power and processes involved in creating inequalities of class, faction, and gender.

Tanya Chiykowski demonstrates the qualities of bundles in her close study of everyday ceramics from the Southwest. Generally considered less invested with meaning, Trincheras traditional plainwares, the material sources, and steps in manufacture and discard provide a departure from processual accounts of pottery manufacture. At each step of the vessel coming-into-being the material collection, vessel formation, decoration, firing, and discard, Chiykowski's account provides clues to how a simple quotidian pot used for cooking or storing water is itself a bundle of sensory experiences invested with meanings and imbricated identities. Chiykowski riffs on the chaîne opératoire of pottery manufacture. Normally seen as a mechanism through which to build an archaeology of practice in studies of lithic and ceramic technology, Chiykowski's approach is to reverse the gaze. By centering on the material and its transformation, she argues we can see the agency or voice of the pot in the making. In short, the spirit of the pot is an ontological beginning. For the archaeologist who recovers the pot, "the characteristics of materials that solidify in the sherds stay with the object long after it has finished its primary use life" (Chiykowski, this volume). There are distinct echoes of object life histories (Joyce 2003; Mills and Walker 2008) where at each step the clay pot accretes a biography.

The focus on bundles and the insistence on rendering the relationship between the body and the world demonstrate how protest and critique can be made visible. Jessica Santos, in her thoughtful discussion of immigrant participation in the 2006 Immigration Reform protests and Occupy Wall Street Movement in 2011 shows how bodies are meshed alongside signs, cameras, stages, flyers, and banners. In the context of late capitalism where the public sphere is increasingly limited, monitored, and dangerous for undocumented migrants, such protest bundles "enable those who live and work in the margins to elude control mechanisms of the state." Yet the composition of protest bundles is not epiphenomenal as experienced by the subject. As described by Santos, analysis of their respective agential and conscious elements in relation to each other would change as each is given additional properties depending on the circumstances. Here *bundle* plays a bit of the trickster figure—on the one hand giving tools for critical analysis of difference, but on the other, at the same time, normalizing such difference in a way that seems natural.

Human Scale

A third theme that has emerged through the course of this volume is that of experience. The relationship between humans and things re/produces relationships of power far beyond the human scale. Yet, at the same time, it is at the very level of bodily experience that one encounters the effective scale (Crumley 1995) of materiality. While informed by earlier incarnations of phenomenological approaches to ancient and modern landscapes

(Brück 2005; Cummings and Whittle 2004; Tilley 1994, 2004), authors find specific inspiration in Heidegger's phenomenology. For Heidegger, knowledge derives from matter, our bodily encounters with matter, and the experiences that derive from those interactions. While phenomenological approaches have been critiqued in the past for assuming a temporal unity between past and present (see Martin Hall's critique of Tilley, Hall [2000:45]), to explore how humans and objects are enchained, bundled, and networked is to potentially introduce dissonance in a world that would otherwise be configured as ordered. This has found particular resonance in Tim Ingold's *taskscape*—the array of related activities and concomitant social relations. As such, the mundane provides moments of extraordinary insight into the construction of past lives.

To expand on this claim, we can turn to Pierre Bourdieu (1977), Nancy Munn (1986), and Michel DeCerteau (1984), who are all fundamentally concerned with materiality and practice. In *Outline of a Theory of Practice* (1977), Bourdieu shows how objects order humans and in so doing shape practice into social agents. Without objects, the practice of everyday life would lose its purchase. For DeCerteau, objects can be implicated in strategies—the actions of the relatively powerful designed to materialize abstractions of control through the ordering of space and things (DeCerteau 1984:28)—and tactics, the calculated actions of the relatively powerless designed to exploit the interstices of strategies (DeCerteau 1984:37). Take the action of consuming goods. It can be a mechanism through which the relatively powerless are disciplined with a social order, but it can also be the locus of appropriation where improvisation can be found in an individual object's use. Again, in the practice of everyday life, things are not just a pale reflection of action, but their instantiation. At a very grounded level, the material turn demands a shift in our "middle range," for lack of a better term. Style, function, standardization, composition, and taphonomy, to name a few, are all basic categories of analysis that have implicit within them repetitive tasks that act as an index to archaeologist, prototype, maker, and user. For many in this volume, a focus on the experience, embodied and performed, provides the critical departure for asking new questions. Authors do this by recasting objects as bundles. But space, and an object's travel through it, plays an important role in materiality's role in interpretation.

Brittany Fullen's chapter provides a thoughtful response to this challenge as she examines the materialization of Wari expansion in the Andes. How is imperial expansion materialized? These pots traveled throughout the Andes, but they did not move by themselves. Huamanga style carried with it Wari identities. Regional variability in these quotidian vessels defies any taxonomy that might organize the closely packed circuitry of people, places, and things and cast a reflection on intent. Potters reacted and reshaped vessels in light of an increased field of stylistic repertoires, and these repertoires were received by those who would use the wares in traditional ways. For Fullen, the vessels speak to a world in which the instability of styles are indistinguishable from the instability of identities as people negotiate their locality and personhood in the wake of Wari expansion. Here there is an echo of Gilles Deleuze (1994) and the manner in which identities and difference are in a dialectical relationship. For Deleuze, identities are never stable categories, nor is difference derived from those categories. Rather the categories are derived from difference that is "virtual." His idiosyncratic use of virtuality rests in two qualities. First is the superficial experience that is generated from material interactions. Second, virtuality is a potential that is materialized through these interactions, producing an actuality. The world of ideas and metaphors are realized in material objects. If we are to take imperial expansion as a wholesale reshaping of the networks of people and things and the ways in which they circulate, we are left with materials that activate these virtual differences and by doing so create new categories.

Virtuality can be glossed as an object's agency, in which things achieve their ability to act on matter through their stance next to human agents. Şule Can describes in particular depth the manner in which objects activate larger landscapes of difference and inequality. Such are the possibilities for a materiality of *zyaras*, sanctuaries visited by the Nusayri of Southern Turkey. The zyaras, a sacred place and a venerated person, are triggered by the repetitive tasks of Nusayri (Arab Alawite), a Muslim religious minority. Can suggests there is value in showing "how people experience these unstable identities and the ways in which they perform in their everyday lives." For the Nusayri who perform many of the same religious acts as Sunnis, difference is marked through the materiality of shrines. Here Can reminds us that materiality provides an exegesis in a landscape dominated by Sunni Islam and Turkish nationalism. At the same time it stabilizes the identities and reinforces Sunni and Turkish chauvinism to this subordinated group (Can, this volume). Can makes a careful distinction between animism and agency and that shrines can only be located in specific places. As such the power of the logic is very much rooted to the ground upon which it is venerated.

In her chapter, Halona Young-Wolfe uses the example of shicras, construction fill that marks monumental architecture in Late Archaic North Coast Peru. As a system of engineering, these netted structures offer little in the way of extra stability. She states that "while the use of shicra is highly diagnostic of Late Archaic architecture, they were not used in every monumental structure at every site or during every construction phase in any given building's history." Young-Wolfe argues that the monument and netting itself is less important than the embodied practices contained in their manufacture. Through the repeated gestures, solidarities are built, and knowledge is transmitted from one generation to the next. Her aim is to get deeper into what has often been assumed to be a simple technological intervention, driven by efficiency or simple renditions of elite power. Wolfe shows how by thinking about the difficulty, coordination, planning, and contingency shicra materialized the circuits of raw material, objects, and people into a kind of solidarity. Yet we should not forget the reasons why shicras were built or the kinds of relations of power they instantiate.

What I find particularly compelling about the phenomenological approaches described above is that they not only allows us to imagine how things worked but also the kinds of failures people encountered on the way. The busted thumb, the rock that escapes the net, and the pot that bursts in the kiln will each have a story to be told at the human scale, which would ideally have a material record. With creative application of an archaeological imagination, such dissonances allows us to shift perspectives from the object, to the social relation, and to the human scale.

Materiality as Problem Space

David Scott (2002:4) defines a problem space as "an ensemble of questions and answers around which a horizon of identifiable stakes (conceptual, as well as ideological-political stakes) hangs. That is to say, what defines this discursive context are not only the particular problems that get posed as problems as such (the problem of 'race,' say), but the particular questions that seems worth asking and the kinds of answers that seem worth having." To extend this idea to the material we can ask to what extent does a focus on materiality help us discern the questions, influences, and inferences to which a prototype, author, and recipient in a particular problem space responded? This should be equally applicable to the past, as well as the present. Additionally, how do objects continue to prompt questions, influences, and inference in new combinations of events and enable archaeologists to determine whether questions are worth asking?

The Palace of Sans Souci in Haiti provides a good illustration (Figure 9.3). The historical anthropologist Michel Rolph Trouillot noted that there

Figure 9.3. The front of the Sans Souci palace in Milot, Haiti. Photo by Rémi Kaupp. Permission was granted to copy, distribute, and/or modify this document under the terms of the GNU Free Documentation.

were three Sans Soucis. There was a palace commissioned by King Henri Christophe of Haiti in 1810 and completed in 1813. There was the Palace of Sans Souci commissioned by Frederick the Great of Prussia, on which the palace in Milot was putatively modeled. There was also the African revolutionary Colonel Jean-Baptist Sans Souci, who continued to fight the French after the submission of Toussait, Dessaline, and Christophe in 1802. A political rival of King Henry, Christophe murdered him, eventually wrote Sans Souci out of history and constructed the palace near his assassination. The palace, its presence, and references reinforce a traditional narrative of monarchal aspirations and silences the existence and actions of a past revolutionary. For Trouillot, the materiality of the palace regulates whether the question "who is Sans Souci" is worth asking (Trouillot 1995).

The example of Trouillot's mapping of the social relations surrounding Sans Souci reveals the linkage among history and culture, power and objects, and inference and interest. Yet such insights are only available through the presence of fragmentary documents and the dissonance those documents create in relation to a crumbling palace on the north coast of Haiti. We are left with the possibilities resident in material networks, bundles, and the repeated gestures of everyday life. As an archaeologist who studies slavery, I see great promise in materiality as a problem space. That being said, practicing materiality in light of slavery reveals two specters. The first specter is that of the evidence we use and how we use it. The second specter is that of our professional, ethical, and political commitments. Both specters are, admittedly, in dialogue with a kind of boundary making/defining what the intellectual projects of anthropology and archaeology should or should not be.

In his commentary on the 2005 volume *Materiality*, Chris Pinney invoked what he referred to as the Ginzburg problem, that is the intellectual habit to "find evidence in the visual that in fact we have discovered elsewhere" (Pinney 2005:260). The evidence itself is incomplete, fragmentary, and in some way instantiating the relations of power that set their presences/absences into being. As messy fragments they can be positioned into previously existing narratives or explanations. This problem is hardly unique to materiality in practice. Historical archaeology has a habit of using sites to verify maps and documents to explain the presence or absence of things. Yet in its uncritical application such materials can reinforce narratives and undermine the potential of adopting a new problem space, such as materiality.

In what way does practicing materiality enable us to seek out information or set forth new questions that might destabilize grand narratives? In short, where is the messiness between humans and nonhumans, and how does this mess provoke questions? Is it the case that within a particular bundle, such as OWS or shicras, that people and things work in concert? A knot can come undone, a text message autocorrected in an inconvenient way, changing the tenor and course of that particular bundle.

The second specter, and one that I am less optimistic about are the unintended implications for stakeholders of the past. If we are to believe, as Michael Shanks and Christopher Tilley argued twenty-five years ago, that the aim of archaeological discourse should be to disempower political and intellectual elites (Shanks and Tilley 1987:195), then we must take seriously all consequences, intended and unintended, of a theoretical apparatus that could be used to render humans as things. Lynn Meskell has made this point explicitly in her discussion of negative heritage and the unintentional privileging that emerges from dichotomies, such as nature and culture (Meskell 2002). It is unclear whether at this particular juncture networks or bundles have sufficient critical apparatus to describe the ways in which life worked in societies with asymmetrical social relations. Given the constraints of space and my desire to maintain focus on this volume, I cannot expand into an analysis here. Instead I would like to pose a series of questions. In what ways does practicing materiality change, alter, or reframe laws related to cultural and national patrimony? How do we incorporate into a humanistic and phenomenological approach, a systematic and rigorous way to find the unexpected?

Authors in this volume have begun to answer these questions in making materiality material. At the heart, there is a central contradiction to archaeology's relationship with things. For me this relationship is intensified by thinking about slavery. Things are traditionally defined in the academy as property, matter, or residue, but being the substance of time, to borrow from Mintz and Price, "they are called on to act in sentient, articulate, and human ways" (1992:43). This makes them particularly mercurial subjects. What is new are the questions these authors attempt to answer in ways that do not make objects epiphenomenal to culture, history, or the putative social forces that bind them.

References Cited

Arcangeli, Myriam. 2015. *Sherds of History: Domestic Life in Colonial Guadeloupe.* University Press of Florida, Gainesville.

Armstrong, Douglas V. 1990. *The Old Village and the Great House: An Archaeological and Historical Examination of Drax Hall Plantation, St. Ann's Bay, Jamaica.* University of Illinois Press, Bloomington.

Bakhtin, Mikhail M. 1981. *The Dialogic Imagination: Four Essays.* Translated by Michael Holquist. University of Texas Press, Austin.

Boivin, Nicole. 2008. *Material Cultures, Material Minds.* University of Cambridge Press, Cambridge.

Bourdieu, Pierre. 1977. *Outline of a Theory of Practice.* Cambridge University Press, Cambridge.

Brück, Joanna. 2005. Experiencing the Past? The Development of a Phenomenological Archaeology in British Prehistory. *Archaeological Dialogues* 12(1):45–72.

Brumfiel, Elizabeth M. 1992. Distinguished lecture in archeology: Breaking and Entering the Ecosystem—Gender, Class, and Faction Steal the Show. *American Anthropologist* 94(3):551–567.

Callon, Michel, John Law, and Arie Rip. 1986. *Mapping the Dynamics of Science and Technology: Sociology of Science in the Real World.* Macmillan, Basingstoke, Hampshire.

Calvino, Italo. 2006 [1978]. *Invisible Cities,* translated by William Weaver. Houghton Mifflin Harcourt, Orlando.

Crumley, Carole L. 1995. Heterarchy and the Analysis of Complex Societies. *Archeological Papers of the American Anthropological Association* 6(1):1–5.

Cummings, Vicki, and Alasdair Whittle. 2004. *Places of Special Virtue: Megaliths in the Neolithic Landscapes of Wales.* Oxbow Books Limited.

de Certeau, Michel. 1984. *The Practice of Everyday Life.* University of California Press, Berkeley.

Deleuze, Gilles. 1994. *Difference and Repetition*. Columbia University Press, New York.

Elias, Norbert. 1994. *The Civilizing Process*. Translated by E. Jephcott. Blackwell, Oxford.

Gell, Alfred. 1998. *Art and Agency: An Anthropological Theory*. Clarendon Press, Oxford.

Gosden, Chris. 2005. What Do Objects Want? *Journal of Archaeological Method and Theory* 12(3):193–211.

Gosden, Chris, and Yvonne Marshall. 1999. The Cultural Biography of Objects. *World Archaeology* 31(2):169–178.

Graeber, David. 2006. Turning Modes of Production Inside Out or, Why Capitalism is a Transformation of Slavery. *Critique of Anthropology* 26(1):61–85.

Hall, Martin. 2000. *Archaeology and the Modern World: Colonial Transcripts in South Africa and the Chesapeake*. Routledge, London.

Handler, Jerome S., and Fredrick Lange. 1978. *Plantation Slavery in Barbados: An Archaeological and Historical Investigation*. Harvard University Press, Cambridge.

Hauser, Mark W. 2011. Routes and Roots of Empire: Pots, Power, and Slavery in the 18th-Century British Caribbean. *American Anthropologist* 113(3):431–447.

———. 2014. Land, Labor, and Things: Surplus in a New West Indian Colony (1763–1807). *Economic Anthropology* 1(1):49–65.

Hicks, Dan. 2010. The Material-Cultural Turn: Event and Effect. In *The Oxford Handbook of Material Culture Studies*, edited by Dan Hicks and Mary C. Beaudry, pp. 25–98. University Press, Oxford.

Higman, Barry W. 2005. *Plantation Jamaica 1750–1850: Capital and Control in a Colonial Economy*. University of the West Indies Press Kingston, Jamaica.

Hodder, Ian. 2012. *Entangled*. Wiley-Blackwell, New York.

Joyce, Rosemary. 2003. Making Something of Herself: Embodiment in Life and Death at Playa de los Muertos, Honduras. *Cambridge Archaeological Journal* 13(02):248–261.

———. 2012. Life with Things: Archaeology and Materiality. In *Archaeology and Anthropology: Past, Present and Future*, edited by David Shankland, pp. 119–132. Berg, Oxford.

Keane, Webb. 2003. Semiotics and the Social Analysis of Material Things. *Language and Communication* 23:409–425.

Kopytoff, Igor. 1986. The Cultural Biography of Things: Commoditization as Process. In *The Social Life of Things: Commodities in Cultural Perspective*, edited by Arjun Appadurai, pp. 64–91. Cambridge University Press, Cambridge.

Las Casas, Bartolome. 1992. *A Short Account of the Destruction of the Indies*. Penguin, New York.

Latour, Bruno. 1993. *We Have Never Been Modern*. Harvard University Press, Cambridge, Mass.

———. 2005. *Reassembling the Social: An Introduction to Actor-Network-Theory*. Oxford University Press, New York.

Law, John, and John Hassard. 1999. *Actor Network Theory and After*. Blackwell, Oxford and Malden, Massachusetts.

McGuire, Randall H. 2013. Steel Walls and Picket Fences: Rematerializing the US–Mexican Border in Ambos Nogales. *American Anthropologist* 115(3):466–480.

Meskell, Lynn. 2002. Negative Heritage and Past Mastering in Archaeology. *Anthropological Quarterly* 75(3):557–574.

Miller, Daniel. 2005. Materiality: An Introduction. In *Materiality*, edited by Daniel Miller, pp. 1–50. Duke University Press, Durham and London.

Mills, Barbara J., and William H. Walker, (editors). 2008. *Memory Work: Archaeologies of Material Practices*. School for Advanced Research Press, Santa Fe.

Mintz, Sidney. 1985. *Sweetness and Power: The Place of Sugar in Modern History*. Viking Press, New York.

Mintz, Sidney W, and Richard Price. 1992 *The Birth of African American Culture: An Anthropological Approach*. Beacon Press, Boston.

Munn, Nancy D. 1986. *The Fame of Gawa: A Symbolic Study of Value Transformation in a Massim (Papua New Guinea) Society*. Cambridge University Press.

Olsen, Bjørnar. 2010. *In Defense of Things: Archaeology and the Ontology of Objects*. Altamira Press, Lanham, Maryland.

Pinney, Christopher. 2005. Things Happen: Or, From Which Moment Does That Object Come? In *Materiality*, edited by Daniel Miller, pp. 256–272. Duke University Press, Durham and London.

Scott, David. 2004. *Conscripts of Modernity: The Tragedy of Colonial Enlightenment*. Duke University Press, Durham and London.

Shanks, Michael, and Christopher Tilley. 1987. *Re-Constructing Archaeology: Theory and Practice*. Cambridge University Press, Cambridge.

Shanks, Michael. 2007. Symmetrical Archaeology. *World Archaeology* 39(4):589–596.

Trouillot, Michel Rolph. 1995. *Silencing the Past: Power and the Production of History*. Beacon Press, New York.

Tilley, Christopher. 1994. *A Phenomenology of Landscape*. Berg, London.

——. 2004. *The Materiality of Stone: Explorations in Landscape Phenomenology*. Oxford University Press, London.

Webmoor, Timothy, and Christopher L. Witmore. 2008. Things Are Us! A Commentary on Human/Things Relations Under the Banner of a "Social" Archaeology. *Norwegian Archaeological Review* 41(1):53–70.

Wilkie, Laurie A. 2000. *Creating Freedom: Material Culture and African American Identity at Oakley Plantation, Louisiana, 1840–1950*. Louisiana State University Press, Baton Rouge.

Witmore, Christopher L. 2007. Symmetrical Archaeology: Excerpts of a Manifesto. *World Archaeology* 39(4):546–562.

Woodward, Robyn P. 2011. Feudalism or Agrarian Capitalism? The Archaeology of the Early Sixteenth-Century Spanish Sugar Industry. In *Out of Many, One People: The Historical Archaeology of Colonial Jamaica*, edited by James A. Delle and Mark W. Hauser, pp. 23–40. University of Alabama Press, Tuscaloosa.

Contributors

Şule Can is a PhD candidate in sociocultural anthropology at Binghamton University – State University of New York (SUNY). She received her BA in English language teaching and literature in 2008 and her MA in cultural studies in 2011 from Istanbul, Turkey. She was awarded a Fulbright scholarship for 2011–2013. She is currently working on a dissertation that focuses on Turkish-Syrian border cities and frontiers. Her research particularly aims to analyze the impact of the Syrian Civil War on Arab minorities in Turkey and contribute to a deeper understanding of state violence, ethno-religious conflicts, and partition in urban space. She is also a columnist and blogger in Turkey who writes about political activism.

Tanya Chiykowski is a PhD candidate in anthropology at Binghamton University – SUNY. She received her BSc in archaeology (Honors) from University of Calgary in 2009, and her MA in anthropology from Binghamton University – SUNY in 2011. Her MA thesis examined the shift in architectural style and domestic space in Chihuahua circa AD 1000. She is the recipient of a National Science Foundation Doctoral Dissertation Improvement Grant to pursue research on technological changes to Sonoran ceramics in the late prehistoric period. The dissertation will assess the impact of trade, migration, and captives on the introduction foreign manufacturing methods to the site of Cerro de Trincheras. Her conference presentations and research interests incorporate a broad range of material culture from Northwest Mexico, including lithic production, pithouse architecture, and ceramic petrography.

Rui Gomes Coelho is an archaeologist interested in historical archaeology, critical theory, and activism. He graduated in 2005 from the New University of Lisbon, Portugal, and in 2010 completed his master's degree at the same university. He has been collaborating with projects based in Portugal, Brazil, Germany, Spain, and the United States and is currently a PhD candidate at Binghamton University – SUNY with the support of Fulbright and the Portuguese Foundation for Science and Technology. His dissertation project focuses on the constitution of the Paraíba Valley (Rio de Janeiro, Brazil) as a landscape of inequality and slavery during the nineteenth century.

Brittany Fullen is a PhD student at Binghamton University – SUNY. She received her BA in anthropology and classical archaeology from the University of Illinois at Urbana-Champaign in 2008 and her MA in anthropology from Binghamton University – SUNY in 2013. Her thesis focused on how Middle Horizon Wari quotidian ceramics were involved in the imperial expansion process and the inherent negotiation of changing identities. Her dissertation focuses on understanding collapse from the perspective of the provincial context in Huari. This project will explore how identities and social relationships were re-negotiated by Wari colonists when the imperial administrative system broke down.

Erina Gruner is completing her doctoral dissertation at Binghamton University – SUNY. Her MA thesis investigated the burial of the Magician from a twelfth century Sinagua site near Flagstaff, Arizona. Her PhD research addresses social organization and exotic exchange in the American Southwest with a focus on Chacoan and post-Chacoan communities. She works for Statistical Research Inc., in Tucson Arizona, and has previously worked for Petrified Forest National Park and Aztec Archaeological Consultants. Gruner is the winner of the 2012 Arizona Archaeological and Historical Society Julian H. Hayden Award for her paper "Re-envisioning Nativism: The Use of Ecclesiastical Paraphernalia During the Pueblo Revolt," published in *Kiva* (2013). She is also the second place winner of the Linda Cordell Prize for her paper "Curating Ancestry: The Afterlives of People and Things at Chacoan and Post-Chacoan Centers," presented at the 2014 Pecos Conference in Blanding, Utah.

Mark W. Hauser is an archaeologist and historical anthropologist who studies how people adapt to landscapes of inequality and contribute to those landscapes in material ways. He employs ethnohistorical, archaeological, and archaeometric approaches to examine the material record of slavery and the social and intellectual contributions of Africans in the New World. He is the author of *An Archaeology of Black Markets* and has edited multiple volumes, including *Islands at the Crossroads* and *Out of Many, One*. He is Associate Professor of Anthropology at Northwestern University.

Jessica Santos López, as a BA student at the University of Puerto Rico (UPR), was a research assistant on a migration project resulting from a research agreement between the anthropology program at the UPR and the Center for Research and Advanced Studies in Social Anthropology, Chiapas. The research explored the reasons why many women leave their communities and resulted in the publication of *Mujeres indígenas, empoderamiento y migración hacia San Cristóbal de Las Casas, Chiapas* (2004). As an MA student at the University of Amsterdam, Santos examined the regimes of truth production that inform U.S. migratory policies, and how nongovernmental and grass-roots organizations contest the discourses produced around the figure of the Latin American immigrant. Currently, she is a PhD candidate and Clark Fellow at Binghamton University – SUNY where she investigates performance practices, how goals are framed and reproduced in activist groups, and how they lead to collective events, such as protests.

Ruth M. Van Dyke (PhD Arizona 1998), volume editor, is Professor of Anthropology at Binghamton University – SUNY. Her archaeological research employs phenomenological and spatial methods to investigate the intersections of memory, materiality, and ideology. Much of her work has been focused on the role of visual and spatial experience in the rise and decline of ancient Pueblo ritual and power at Chaco Canyon in the Southwest United States. She is the author of *The Chaco Experience: Landscape and Ideology at the Center Place* (2008), senior editor of *Archaeologies of Memory* (with Susan Alcock, 2003), and author of approximately forty additional articles and chapters on the archaeology of the ancient Southwest United States. As senior editor of *Subjects and Narratives in Archaeology* (with Reinhard Bernbeck, 2015), she is exploring alternative forms of archaeological narration. In addition to ongoing work at Chaco, she directs a historical archaeology project investigating bodies, spaces, and objects in the construction of ethnic identities in nineteenth-century Texas. She is currently at work on an interdisciplinary study of the materiality of pilgrimage.

Halona Young-Wolfe earned her BA in anthropology and English from Drew University in 1996 and her MA in anthropology from Binghamton University – SUNY in 2013. She has worked at the Late Archaic site of

Aspero in Peru and at the Shell Mound site in Florida. Young-Wolfe is interested in questions of social complexity and the emergence of inequality. Her dissertation research at Binghamton University – SUNY focuses on domestic architecture from the Late Archaic period in the Central Andes and investigates the evidence for social hierarchy at the household level.

Index

abduction. *See under* Gell, Alfred

actor-network theory (ANT), 9–10, 18, 56. *See also* Latour, Bruno

American Southwest, 11, 17, 23, 56–75, 79–96, 198, 202, 204

Andean South America, 22–24, 45, 124–44, 149–72, 206

animacy, and objects, 17, 23, 45, 51, 57–59, 65, 73, 79–89, 94–96, 101, 202, 207. *See also* animals; ceramic vessels, effigy

animals, 5, 7, 15–18, 24, 45, 65, 69, 106, 125, 130, 155–58. *See also* animacy; ceramic vessels, effigy; macaws; parrots

animism. *See* animacy; animals

anthropocentrism, 7, 18–20, 197–98, 204

Appadurai, Arjun, 8, 81

Arab Alawite (Nusayri), 17, 21, 33–54

architecture, monumental, 23, 66–70, 73, 84, 117, 149, 159–61, 164–65, 169–72, 207, 208. *See also* buildings and houses; cities; places

assemblages, 3, 4, 6, 7, 11, 12, 15, 20, 21, 23, 25, 56, 58–74, 79, 118, 124–25, 128, 135, 137, 183

basketry, 24, 62, 64, 69, 70, 81, 91, 92, 166, 167, 169

Bennett, Jane, 16, 18–19. *See also* vital materialism

bodies: as clay or vessels, 86, 89, 105–6, 108, 130, 179, 203; of the dead, 11, 39, 51, 67, 74, 88, 94, 101, 109; practices and, 13, 24, 41–46, 51, 65, 151–52,

155–59, 166–70, 180, 186, 191, 196–98, 206; sensory experiences and, 4, 13, 15–17, 21–22, 43–51, 91, 119, 154–58, 170, 177–79, 196–98, 205–6; the state and, 22, 24, 100, 176–93, 205. *See also* Foucault, Michel; performance

Bourdieu, Pierre: and *habitus*, 5, 181, 206; and practice, 5, 206

buildings and houses, 21, 33, 39, 42–44, 46, 50, 84, 89, 100, 104, 106, 109, 110, 112, 114, 115, 119, 160, 184; construction of, 10, 40, 41, 67, 70, 72, 96, 101, 103, 150, 160–70, 207–8; destruction of, 84, 103, 109, 112, 115. *See also* architecture, monumental; cities

bundles, 6–7, 11–12, 25, 51, 56–57, 100, 105, 107, 128, 204, 206, 209–10; of construction material, 22, 170; medicine, 11, 17, 57; of ritual objects, 22–23, 51, 57, 61–75, 96, 202, 204, pots as part of, 22, 79, 81, 87, 93, 96, 204–5; protest, 22, 24, 176–93, 205. *See also* ritual paraphernalia; *shicra*

burials. *See* bodies, of the dead; objects, mortuary

Caribbean, 24, 198–200, 203, 208–9

Cartesian dualisms, 4, 6, 16, 18, 52, 150, 157–58, 179

ceramic vessels: discard of, 23, 81, 88, 93–95, 204–5; effigy, 79, 86, 87, 92, 95, 130, 131, 135, 136–38, 141;

219